Lecture Notes in Computer Science 6613

Commenced Publication in 1973
Founding and Former Series Editors:
Gerhard Goos, Juris Hartmanis, and Jan van Leeuwen

Editorial Board

David Hutchison
  Lancaster University, UK
Takeo Kanade
  Carnegie Mellon University, Pittsburgh, PA, USA
Josef Kittler
  University of Surrey, Guildford, UK
Jon M. Kleinberg
  Cornell University, Ithaca, NY, USA
Alfred Kobsa
  University of California, Irvine, CA, USA
Friedemann Mattern
  ETH Zurich, Switzerland
John C. Mitchell
  Stanford University, CA, USA
Moni Naor
  Weizmann Institute of Science, Rehovot, Israel
Oscar Nierstrasz
  University of Bern, Switzerland
C. Pandu Rangan
  Indian Institute of Technology, Madras, India
Bernhard Steffen
  TU Dortmund University, Germany
Madhu Sudan
  Microsoft Research, Cambridge, MA, USA
Demetri Terzopoulos
  University of California, Los Angeles, CA, USA
Doug Tygar
  University of California, Berkeley, CA, USA
Gerhard Weikum
  Max Planck Institute for Informatics, Saarbruecken, Germany

Jordi Domingo-Pascual   Yuval Shavitt
Steve Uhlig (Eds.)

Traffic Monitoring and Analysis

Third International Workshop, TMA 2011
Vienna, Austria, April 27, 2011
Proceedings

 Springer

Volume Editors

Jordi Domingo-Pascual
Universitat Politècnica de Catalunya (UPC) - Barcelona TECH
Departament d'Arquitectura de Computadors, Campus Nord. Mòdul D6
Jordi Girona 1-3, 08034 Barcelona, Spain
E-mail: jordi.domingo@ac.upc.edu

Yuval Shavitt
Tel Aviv University, The Iby and Aladar Fleischman Faculty of Engineering
School of Electrical Engineering, Wolfson Building for Software Engineering
Ramat Aviv 69978, Israel
E-mail: shavitt@eng.tau.ac.il

Steve Uhlig
Technische Universität Berlin, An-Institut Deutsche Telekom Laboratories
FG INET, Research Group Anja Feldmann, Sekr. TEL 16
Ernst-Reuter-Platz 7, 10587 Berlin, Germany
E-mail: steve@net.t-labs.tu-berlin.de

ISSN 0302-9743   e-ISSN 1611-3349
ISBN 978-3-642-20304-6   ISBN 978-3-642-20305-3 (eBook)
DOI 10.1007/978-3-642-20305-3
Springer Heidelberg Dordrecht London New York

Library of Congress Control Number: 2011924344

CR Subject Classification (1998): C.2, D.4.4, H.3, H.4, D.2

LNCS Sublibrary: SL 5 – Computer Communication Networks and Telecommunications

© Springer-Verlag Berlin Heidelberg 2011
This work is subject to copyright. All rights are reserved, whether the whole or part of the material is
concerned, specifically the rights of translation, reprinting, re-use of illustrations, recitation, broadcasting,
reproduction on microfilms or in any other way, and storage in data banks. Duplication of this publication
or parts thereof is permitted only under the provisions of the German Copyright Law of September 9, 1965,
in its current version, and permission for use must always be obtained from Springer. Violations are liable
to prosecution under the German Copyright Law.
The use of general descriptive names, registered names, trademarks, etc. in this publication does not imply,
even in the absence of a specific statement, that such names are exempt from the relevant protective laws
and regulations and therefore free for general use.

Typesetting: Camera-ready by author, data conversion by Scientific Publishing Services, Chennai, India

Printed on acid-free paper

Springer is part of Springer Science+Business Media (www.springer.com)

Preface

The Third International Workshop on Traffic Monitoring and Analysis (TMA 2011) was an initiative of the COST Action IC0703 "Data Traffic Monitoring and Analysis: Theory, Techniques, Tools and Applications for the Future Networks" http://www.tma-portal.eu/cost-tma-action.

The COST program is an intergovernmental framework for European Cooperation in Science and Technology, promoting the coordination of nationally funded research on a European level. Each COST Action aims at reducing the fragmentation in research and opening the European research area to cooperation worldwide.

Traffic monitoring and analysis (TMA) is nowadays an important research topic within the field of computer networks. It involves many research groups worldwide that are collectively advancing our understanding of the Internet. Modern packet networks are highly complex and ever-evolving objects. Understanding, developing and managing such infrastructures is difficult and expensive in practice. Traffic monitoring is a key methodology for understanding telecommunication technology and improving its operation, and TMA-based techniques can play a key role in the operation of real networks. Besides its practical importance, TMA is an attractive research topic for several reasons. First, the inherent complexity of the Internet has attracted many researchers to face traffic measurements since the pioneering times. Second, TMA offers a fertile ground for theoretical and cross-disciplinary research, such as the various analysis techniques being imported into TMA from other fields, while at the same time providing a clear perspective for the practical exploitation of the results. In other words, TMA research has the potential to reconcile theoretical investigations with real-world applications, and to realign curiosity-driven with problem-driven research.

In the spirit of the COST program, the COST-TMA Action was launched in 2008 to promote building a research community in the specific field of TMA. Today, it involves research groups from academic and industrial organizations from 25 countries in Europe.

The goal of the TMA workshops is to open the COST Action research and discussions to the worldwide community of researchers working in this field. Following the success of the first two editions of the TMA workshop in 2009 and 2010, we decided to maintain the same format for this third edition: single-track full-day program. TMA 2011 was organized jointly with the European Wireless conference and was held in Vienna on April 27, 2011.

TMA 2011 attracted 29 submissions. Each paper was carefully reviewed by at least three members of the Technical Program Committee. The reviewing process led to the acceptance of ten papers as full papers and four as short papers. Short papers are works that provide promising results or justify discussion during the workshop, but lack technical maturity to be accepted as full papers. Finally, this year's workshop included a poster session. This poster session welcomes work-in-progress work from academia as well as industry. Early experimental results, insights, and prototypes are also of interest. Six posters were accepted.

We would like to thank all members of the Technical Program Committee for their timely and thorough reviews. We hope you will enjoy the workshop and make this event a success!

April 2011

Jordi Domingo-Pascual
Yuval Shavitt
Steve Uhlig

Organization

Program Committee Chairs

Jordi Domingo-Pascual	Technical University of Catalonia (UPC), Spain
Yuval Shavitt	Tel Aviv University, Israel
Steve Uhlig	TU Berlin/T-labs, Germany

Program Committee

Pere Barlet-Ros	Technical University of Catalonia (UPC), Spain
Ernst Biersack	Eurécom, France
Pierre Borgnat	ENS Lyon, France
Albert Cabellos	Technical University of Catalonia (UPC), Spain
Kenjiro Cho	IIJ, Japan
Ana Paula Couto da Silva	Federal University of Juiz de Fora, Brazil
Xenofontas Dimitropoulos	ETH Zürich, Switzerland
Benoit Donnet	Université catholique de Louvain (UCL), Belgium
Gianluca Iannaccone	Intel Research Berkeley, USA
Ethan Katz-Bassett	University of Washington, USA
Udo Krieger	University of Bamberg, Germany
Youngseok Lee	CNU, Korea
Olaf Maennel	Loughborough University, UK
Gregor Maier	ICSI, Berkeley, USA
Marco Mellia	Politecnico di Torino, Italy
Michela Meo	Politecnico di Torino, Italy
Philippe Owezarski	CNRS, Toulouse, France
Dina Papagiannaki	Intel Pittsburgh, USA
Antonio Pescape	University of Naples, Italy
Aiko Pras	University of Twente, The Netherlands
Fabio Ricciato	University of Salento, Italy and FTW, Austria
Dario Rossi	TELECOM ParisTech, France
Luca Salgarelli	University of Brescia, Italy
Ruben Torres	Purdue University, USA
Tanja Szeby	Fraunhofer FOKUS Berlin, Germany

Local Arrangements Chairs

Fabio Ricciato   University of Salento, Italy and FTW, Austria
Alessandro D'Alconzo   FTW, Austria

Poster Session Committee

Saverio Niccolini   NEC Laboratories Europe, Germany
Giuseppe Bianchi   Univeristà di Roma 2, Italy
Brian Trammell   ETH Zürich, Switzerland
Nuno M. Garcia   Universidade Lusófona de Humanidades e
                              Tecnologias, Portugal
Chadi Barakat   INRIA, Sophia Antipolis, France

Table of Contents

Traffic Classification

Poster Session

On Profiling Residential Customers

Marcin Pietrzyk[1], Louis Plissonneau[1],
Guillaume Urvoy-Keller[2], and Taoufik En-Najjary[1]

[1] Orange Labs, France
{marcin.pietrzyk,louis.plissonneau,
taoufik.ennajjary}@orange-ftgroup.com
[2] Université de Nice Sophia-Antipolis, Laboratoire I3S CNRS UMR 6070, France
urvoy@unice.fr

Abstract. Some recent large scale studies on residential networks (ADSL and
FTTH) have provided important insights concerning the set of applications used
in such networks. For instance, it is now apparent that Web based traffic is dominating again at the expense of P2P traffic in lots of countries due to the surge
of HTTP streaming and possibly social networks. In this paper we confront the
analysis of the overall (high level) traffic characteristics of the residential network
with the study of the users traffic profiles. We propose approaches to tackle those
issues and illustrate them with traces from an ADSL platform. Our main findings
are that even if P2P still dominates the first heavy hitters, the democratization of
Web and Streaming traffic is the main cause of the come-back of HTTP. Moreover, the mixture of applications study highlights that these two classes (P2P
vs. Web + Streaming) are almost never used simultaneously by our residential
customers.

1   Introduction

The research community has devoted significant efforts to profile residential traffic in
the last couple of years. A large scale study of Japanese residential traffic [5,4], where
almost 40% of the Internet traffic of the island is continuously observed, has revealed
specific characteristics of the Japanese traffic: a heavy use of dynamic ports, which
suggests a heavy use of P2P applications and a trend of users switching from ADSL to
FTTH technology to run P2P along with gaming applications. A recent study in the US
[6], where the traffic of 100K DSL users has been profiled with a Deep Packet Inspection tool, has revealed that HTTP traffic is now the dominant protocol at the expense of
P2P for the considered ISP, and probably for the US in general. This significant change
in the traffic breakdown is not due to a decrease of P2P traffic intensity but a surge of
HTTP traffic driven by HTTP streaming services like YouTube and Dailymotion. Similar results have been obtained in European countries. In Germany, a recent study [11]
analyzed about 30K ADSL users and also observed that HTTP was again dominant at
the expense of P2P traffic, for the same reason as in the US: a surge of video content
distribution over HTTP. Early studies in France [15] for an ADSL platform of about
4000 users highlighted the dominance of P2P traffic in 2007 but a subsequent studies

J. Domingo-Pascual, Y. Shavitt, and S. Uhlig (Eds.): TMA 2011, LNCS 6613, pp. 1–14, 2011.
© Springer-Verlag Berlin Heidelberg 2011

on the same PoP [14] or other PoPs under the control of the same ISP revealed similar traffic trend of HTTP traffic increasing at the expense of P2P both for ADSL [12] and FTTH access technology [17]. In the above studies, the application profiling of residential traffic was used to inform network level performance aspects, e.g., cachability [6] of content or location in the protocol stack of the bottleneck of transfers performed on ADSL networks [15], [11]. The study in [11] further reports on usage of the ADSL lines with a study of the duration of Radius sessions.

The current work aims at filling the gap between the low-level (network) level performance study and high level (application) study by profiling ADSL users. We use hierarchical clustering techniques to aggregate users' profiles according to their application mix. Whereas many studies focus on communication profiles on backbone links, few ones dig into application mix at user level. In the analysis carried in [10], the authors take a *graphlet* approach to profile end-host systems based on their transport-layer behavior, seeking users clusters and "significant" nodes. Authors in [8], take advantage of another clustering technique (namely Kohonen Self-Organizing Maps) to infer customers application profiles and correlate them with other variables (*e.g.* geographical location, customer age).

Our raw material consists of two packet traces collected on the same platform, a few months apart from each other, that are fully indexed in the sense that both IP to user and connection to applications mapping are available. We use this data to discuss different options to profile both a platform and the users of this platform.

The remaining of this paper is organized as follows. In Sect. 2, we detail our data sets. In Sect. 3, we analyze high level traffic characteristics and, the contributions of users to the traffic per application. In Sect. 4, we discuss different options to profile users and come up with a specific approach that allows to understand application usage profiles.

2   Data Set

The raw data for our study consists of two packet level traces collected on an ADSL platform of a major ISP in France (Tab. 1). Each trace lasts one hour and aggregates all the traffic flowing in and out of the platform.

In this platform, ATM is used and each user is mapped to a unique pair of Virtual Path, Virtual Channel, identifiers. As the packet level trace incorporates layer 2 information, we can identify users thanks to this ATM layer information. This approach allows for reliable users tracking. Indeed, 18% of the users change their IP address at

Table 1. Traces summary

Label	Start time	Duration	Bytes	Flows	TCP Bytes	TCP Flows	Local Users	Local IPs	Distant IPs
Set A	2009-03-24 10:53 (CET)	1h	31.7G	501K	97.2 %	30.7 %	1819	2223	342K
Set B	2009-09-09 18:20 (CET)	1h	41 G	796K	93.2 %	18.3 %	1820	2098	488K

least once, with a peak at 9 for one specific user. One could expect that the only source of error made when considering the IP address is that the session of the user is split onto several IP level sessions. However, we also noticed in our traces that a given IP could be reassigned to different users during the periods of observation. Specifically, 3% of the IPs were assigned to more than one user, with a peak of 18 re-assignments for one specific IP. Those results are in line with the ones obtained in [11] for a German residential operator.

Both traces are indexed thanks to a Deep Packet Inspection (DPI) tool developed internally by the ISP we consider. This tool is called ODT. In [13], we have compared ODT to Tstat (`http://tstat.tlc.polito.it/`), whose latest version features DPI functions. Specifically, we have shown that ODT and Tstat v2 offer similar performance (for most popular applications) and outperform signature based tools used in the literature. As ODT embeds a larger set of signatures than Tstat v2, we rely on the former to map flows and applications.

The classes of traffic we use along with the corresponding applications are reported in Tab. 2. Note that HTTP traffic is broken into several classes depending on the application implemented on top: Webmail is categorized as mail, HTTP streaming as streaming, HTTP file transfers as DOWNLOAD, etc. The OTHERS class aggregates less popular applications that ODT recognized. The DOWNLOAD class consists mainly of HTTP large file transfers from one-click hosting services [1], which are growing competitors of P2P file sharing services. The flows not classified by ODT (e.g. some encrypted applications) are aggregated in the UNKNOWN class.

We developed an ad-hoc C++ trace parser that relies on libpcap to extract the per user statistics from the raw traces. Users' data was anonymized prior to analysis.

Table 2. Application classes

Class	Application/protocol
WEB	HTTP and HTTPs browsing
UNKNOWN	–
P2P	eDonkey, eMule obfuscated, Bittorrent Gnutella, Ares, Others
MAIL	SMTP, POP3, IMAP, IMAPs POP3s, HTTP Mail
CHAT	MSN, IRC, Jabber Yahoo Msn, HTTP Chat
STREAMING	HTTP Streaming, Ms. Media Server, iTunes, Quick Time
OTHERS	NBS, Ms-ds, Epmap, Attacks
DB	LDAP, Microsoft SQL, Oracle SQL, mySQL
DOWNLOADS	HTTP file transfer, Ftp-data, Ftp control
GAMES	NFS3, Blizzard Battlenet, Quake II/III Counter Strike, HTTP Games
VOIP	Skype
NEWS	Nntp

Table 3. Traffic Breakdown (Classes with more than 1% of bytes only)

	Set A	Set B
Class	Bytes	Bytes
WEB	22.68 %	20.67 %
P2P	37.84 %	28.69 %
STREAMING	25.9 %	24.91 %
DOWNLOAD	4.31 %	6.47 %
MAIL	1.45 %	0.54 %
OTHERS	1.04 %	0.44 %
VOIP	0.36 %	1.67 %
UNKNOWN	5.26 %	15.79 %

3   Platform Profile

In this section, we highlight high level platform traffic profiles, namely the traffic breakdown and the per users volume distribution.

3.1   Traffic Breakdown

We report in Tab. 3 the bytes breakdown views of the two traces, where the DB, CONTROL, NEWS, CHAT and GAMES classes have been omitted as they do not represent more than 1% of bytes and flows in any of the traces. It has been observed in [6] and [11] that HTTP based traffic was again dominating at the expense of P2P traffic in residential networks in US and Europe. The traffic breakdown of our platform suggests the same conclusion. Indeed, when summing all HTTP-based traffic in sets A or B, namely Web, HTTP Streaming and HTTP Download, more than 50% of the bytes in the down direction is carried over HTTP. Clearly, HTTP driven traffic dominates at the expense of background traffic that is due to P2P applications.

3.2   Distributions of Volumes per User

Understanding the relative contribution of each user to the total amount of bytes generated by dominating applications is important. Indeed these results, even if not surprising, justify the approach of focusing on heavy hitters[1] in the last section of the paper.

In Fig. 1, we present the contribution of users to the total traffic aggregate per application, with users sorted by decreasing volumes for the considered application (sets A and B being similar we focus on set A here). Note that we sum up, for a user, her bytes in both directions. We also include in the graph the overall contribution by user without distinguishing per application.

The fraction of users contributing to the majority of bytes in each application and even overall is fairly small. When looking at the global volumes generated, 90% of the

---

[1] We term as *heavy hitter* user that is responsible for large fraction of bytes transfered on the platform.

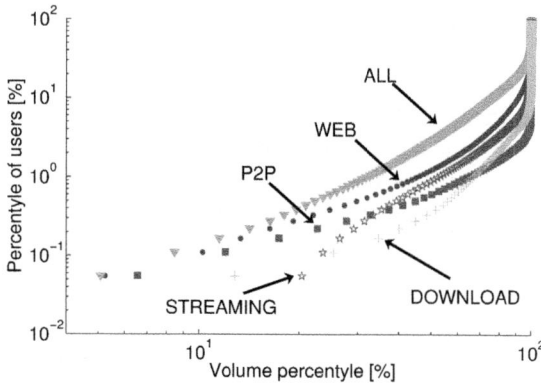

Fig. 1. Contribution of users to traffic aggregate (global and per application). Set A.

bytes are generated by about 18% of users. For the same volume quantile, the fraction of users involved is even smaller when focusing on the applications generating most of the bytes (those represented in the graph). For the case of P2P traffic for instance, only 0.3% of the users contribute to 90% of the bytes uploaded or downloaded. We confirm here the well known phenomenon explored for instance in [7,3]. This also holds for the Streaming and Web classes, which are two key classes in the dimensioning process of links of ISPs (for example bulk of Streaming users is active in the evenings).

A consequence of these highly skewed distributions is that the arrival or departure of some customers on the platform can potentially have an important impact on the traffic shape. For instance, the first four heavy users of streaming are responsible for about 30% of all streaming traffic.

The above observations also motivates our approach in the next section which is on profiling customers (and especially heavy hitters) from their application usage perspective.

4   Users Profiling

In this section, we address the issue of building an application level profile of customers that would characterize their network usage. The problem is challenging as it can be addressed from many different viewpoints. Here are some questions that one might want to answer: Which amount of bytes or alternatively which number of flows should be observed to declare that a user is actually using a specific application? Can we characterize users thanks to the dominant application they use? What is the typical application profile of a heavy hitter? What is the typical application mix of the users?

We address the above questions in the next paragraphs. We discuss several options to map applications to users. Our first approach focuses on the dominating applications for each user, we further discuss the precise profile of the top ten heavy hitters in both traces. Last paragraph presents typical users application mixture using clustering technique.

4.1   Users Dominating Application

We present here a simple approach that provides an intuitive high level overview of the users activity: we label each user with her dominating application, the application that generated the largest fraction of bytes. Such an approach is justified by the fact that for both of our data sets, the dominating application explains a significant fraction of the bytes of the user. Indeed, for over 75% of the users, it explains more than half of the bytes. This phenomenon is even more pronounced when considering heavy users. Fig. 2 presents the distribution of the fraction of the bytes explained depending on which application dominates users activity.

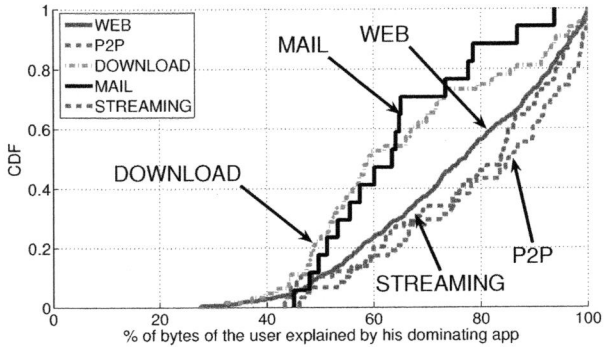

Fig. 2. CDF of the fraction of bytes explained by the dominant application of each user. Set B.

The distribution of users per application with such an approach (dominant application) is reported in Tab. 4. As expected, the dominating class is Web. We have more Streaming than P2P dominated users. This complies with the intuition that every user, even if not experienced, can watch a YouTube video, whereas using a P2P application requires installing a specific software (P2P client). The remaining dominant applications correspond to clients that generate a small amount of bytes most of the time. For instance, users that have DB, Others, Control or Games as dominating application generate an overall number of bytes that is extremely low.

We present in Fig. 3 the users to application mapping for set B using the above dominant application approach. We adopt a representation in which each user is characterized by the total number of bytes she generates in the up and down direction and label the corresponding point in a two dimensional space with the dominant application of the user in terms of bytes. We restricted the figure to a list of 6 important applications: Web, Streaming, VOIP, Download and P2P. We further added the users having majority of bytes in the Unknown class to assess their behavior.

Most important lesson of Fig. 2 is that labeling a client with her dominating application is meaningful. Indeed, the dominating application in terms of bytes usually generates the vast majority of users' total volume. Customers with the same dominating applications are clustered together, and exhibit behavior typical for this application, which we detail below.

Table 4. Users dominating applications breakdown. Each user is labeled with his dominant application in terms of bytes. (Only users that transfered at least 100B: 1755 users). Set B.

Class	Fraction of Users	Fraction of Bytes explained
UNKNOWN	21%	12%
WEB	35%	19%
P2P	4%	35%
DOWN	5%	$\leq 1\%$
MAIL	1%	$\leq 1\%$
DB	9%	$\leq 1\%$
OTHERS	8%	$\leq 1\%$
CONTROL	7%	$\leq 1\%$
GAMES	$\leq 1\%$	$\leq 1\%$
STREAMING	7%	25%
CHAT	1%	$\leq 1\%$
VOIP	1%	2%

We observe from Fig. 3 that:

- P2P heavy hitters tend to generate more symmetric traffic than Download and Streaming heavy hitters, which are far below the bisector.
- Web users fall mostly in between the bisector and the heavy hitters from the Download and Streaming classes. This is also in accordance with intuition as Web browsing often requires data exchange from clients to servers, e.g., when using Web search engines. This is in contrast to Streaming or Download where data flow mainly from servers to clients.

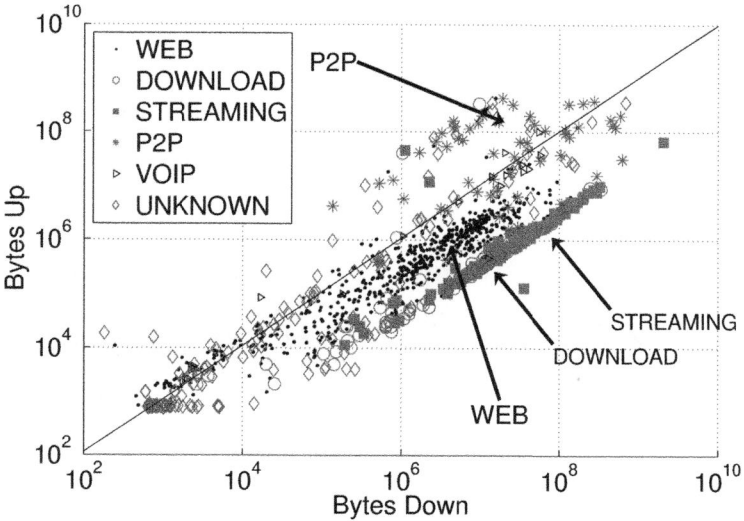

Fig. 3. Users bytes Up/Down. Dominating application marked. Set B.

– Concerning Unknown users, we observe first that a significant fraction of them generated almost no traffic as they lay in the bottom-left corner of the plot. As for Unknown heavy hitters, we observe that they are closer on the figure to P2P heavy users than to client-server heavy users. This might indicate that there exist some P2P applications that fly below the radar of our DPI tool. We further investigate this issue in the next section.

A last key remark is that the equivalent of Fig. 3 for set A is qualitatively very similar, emphasizing the overall similarity of users activity in the two data sets (even if several month apart and at a different time of day).

The above analysis has again underlined the crucial role of (per application) heavy hitters. In the next section, we will focus on the top 10 heavy hitters in each trace. Each of them generated at least 0.6 GB of data and up to 2.1 GB and, overall, they are responsible for at least 1/4 of the bytes in each trace. We profile these users by accounting simultaneously for all the applications they use.

4.2   Top Ten Heavy Hitters

In this section, we focus on the top 10 heavy hitters for sets A and B. Note that these are distinct sets of users. It is a small, but very important group of customers from the ISP perspective, and better understanding of this group (aggregating 1/4 of total volume) might have significant impact on network provisioning and dimensioning. Fig. 4(a) and 4(b) show the fraction of bytes they have generated in the up (positive values) and down direction (negative values) for each application. For sake of clarity, we put in the figure only the labels of the significant applications for each user. We do observe from Fig. 4(a) and 4(b) that heavy hitters, for the most part, use P2P applications. Streaming and (at least for one user) download activities seem also to give birth to some heavy hitters.

We also observe that unknown traffic seems to be associated mostly with P2P users (which is in line with Fig. 3). This is an important finding from the perspective of the traffic classification, which often relies on per flow features. This user level information could be used as a feature in the classifier. It is also in line with the findings in [12] where it is shown that a significant fraction of bytes in the unknown category (we use the same DPI tool but different traces) is generated by P2P applications. In the present case, 67 % and 95 % of unknown bytes are generated by the users having in parallel

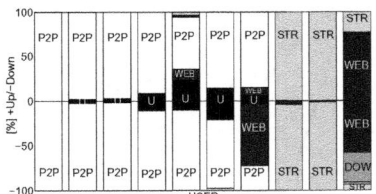 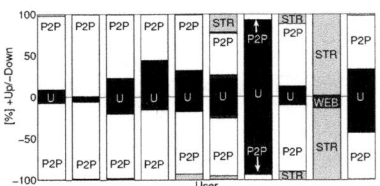

(a) Set A (Users generating 31% of the bytes in the trace)

(b) Set B (Users generating 24% of the bytes in the trace)

Fig. 4. Top 10 heavy hitter users. Application usage profiles expressed in bytes fractions. (U stands for UNKNOWN).

peer-to-peer activity for set A and B respectively. The reason why some of the P2P traffic might be missed by our DPI tool is out of the scope of the paper. We note that there are at least two possible explanations: either we missed in our trace the beginning of a long P2P transfer and the DPI tool might not have enough information[2] to take a decision, or these users run currently unknown P2P applications in parallel.

4.3 Users Application Mix

In the previous sections, we analyzed our users profile taking only bytes into account. This approach is informative and makes sense from a dimensioning viewpoint. However as the per applications volumes are very different – e.g., P2P applications tend to generate much more bytes than Web browsing – we miss some usage information with this purely byte-based approach. In this section, we explore a different perspective. We associate to each user a binary vector, which indicates her usage of each application. We take advantage of clustering techniques to present typical application mixtures.

"Real" *vs.* "fake" usage. We represent each customer with a binary vector: $A = [appli_1 \cdots appli_n]$ where n is the number of applications we consider. Each $appli_i \in \{0, 1\}$ is a indication weather the customer used application i or not. We define per application heuristics to declare that a customer actually uses a class of application. To do that, we define minimal thresholds for three metrics: bytes up, bytes down and number of flows. Depending on the application any or all of the three thresholds need to be matched. We summarize the heuristics in Tab. 5. The values were derived from the data as it is exemplified in Fig. 5 for P2P and WEB traffic.

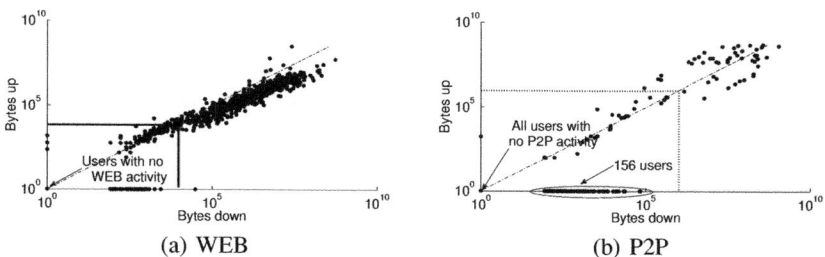

(a) WEB   (b) P2P

Fig. 5. Example of how the threshold is selected. Set A.

Heuristics are necessary to separate real application usage from measurements artifacts (for instance misclassification due to not enough payload). For instance, large fraction of users of the platform have a single flow which is declared by the DPI tool as WEB browsing. It is hard to believe that this flow is a real web browsing activity, as current web sites tend to generate multiple connections for a single site (single search without browsing on `google.com` shows up to 7 connections). Similar problems might occur with other applications, for instance peer-to-peer user that closed his application, might still receive file requests for some time due to the way some P2P overlays work.

---

[2] Application level information are often at the onset of transfers [2].

Table 5. Ad-hoc, per application and user minimum thresholds to declare application usage

Class	Volume Down	Up	Number of Flows	Policy
WEB	300kB	500kB	20	All
P2P	1 MB	1 MB	10	Any
STREAMING	1 MB	1 MB	–	Any
DOWNLOAD	2 kB	1 kB	–	Any
MAIL	30kB	3 kB	–	All
GAMES	5 kB	5 kB	–	Any
VOIP	200kB	200kB	–	All
CHAT	10kB	10kB	–	Any

Choice of clustering. We have considered several popular clustering techniques to be able to understand the application mix of each user, see [9] for a complete reference on main clustering techniques. As explained in the previous paragraph, we have discretized the user's characteristics according to some heuristic threshold in order to keep only "real" application usage.

We have first tried the popular k-means clustering algorithm, and observed that the resulting clusters are difficult to match to applications. Moreover the choice of the number of clusters can dramatically change this representation.

Hierarchical clustering offers an easily interpretable technique for grouping similar users. The approach is to take all the users as tree leaves, and group leaves according to their application usage (binary values). We choose an agglomerative (or down-up) method:

1. The two closest nodes[3] in the tree are grouped together;
2. They are replaced by a new node by a process called linkage;
3. The new set of nodes is aggregated until there is only a single root for the tree.

With this clustering algorithm, the choices of metric and linkage have to be customized for our purpose.

We want to create clusters of users that are relatively close considering the applications mix they use. Among comprehensive metrics for clustering categorical attributes the Tanimoto distance [16] achieves these requirements. It is defined as follows: $d(x,y) = 1 - \frac{x^t \cdot y}{x^t \cdot x + y^t \cdot y - x^t \cdot y}$.[4] This means that users having higher number of common applications will be close to each other. For example, consider 3 users having the following mix of applications[5]:

User	Web	Streaming	Down	P2P
A	1	1	0	0
B	1	1	1	0
C	1	1	0	1

---

[3] At first occurrence, nodes are leaves.

[4] x^t stands for x transposed.

[5] 1 means application usage and 0 means no application usage.

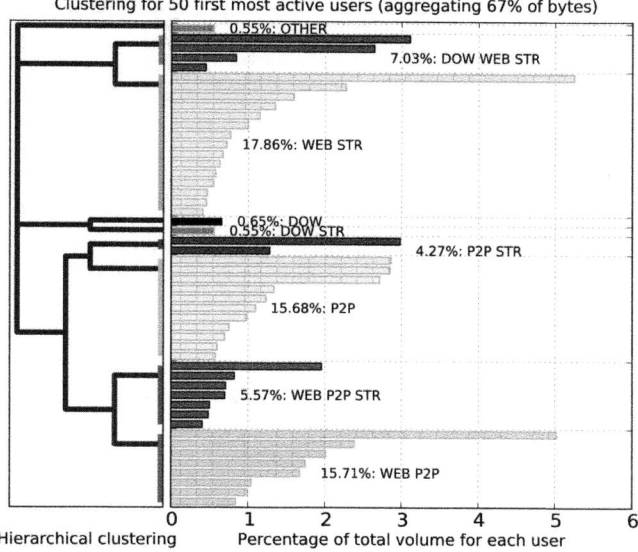

Fig. 6. Application clustering for top 50 most active users. Set A.

With Tanimoto distance, users B and C will be closer to each other because they have same total number of applications even if all 3 users share same common applications.

We use a complete linkage clustering, where the distance between nodes (consisting of one or several leaves) is the maximum distance among every pair of leaves of these nodes. It is also called farthest neighbor linkage.

Due to the chosen metric, and as we chose not to prune the resulting tree, the hierarchical clustering leads to as many clusters as there are applications combinations: $\sum_{i=1}^{n} \binom{n}{i}$. In our case, we restrict the set of applications we focus only to Web, Streaming, P2P and Download.

Applications mix. We present in Fig. 6 and 7 the clustering results for the top 50 and second 50 most active users respectively. In total, the first one hundred users of the platform are responsible for 80% of the volume. We first consider only the classes generating most of the traffic, as described by Tab. 3 namely: Web, P2P, Streaming, and Download.

Each barplot represents a single user and expresses his total volume share. Barplots (thus users) are grouped into the sorted clusters. Each cluster, indicated by a different color, groups the users that had the same applications. Thus close clusters in the graph are similar with respect to their application mix.

Considering only four applications, we have 15 possible combinations. What we observe is that some combinations are clearly more popular than others, while a few of them never occurred in our data. We present below a more precise analysis that reveals some insights about the typical users profiles.

Looking at the top 50 active users, we see that the P2P related clusters (P2P only, P2P + Web, P2P + Web + Streaming) dominates the top heavy hitters with 28 users.

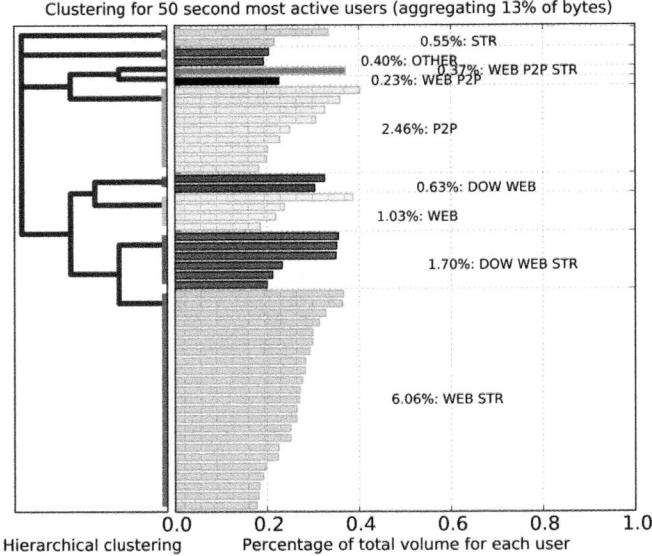

Fig. 7. Application clustering for top 51-100 most active users. Set A.

These P2P related clusters aggregate 36% of the total volume of the trace. Pure Web + Streaming profiles are the largest cluster in volume (18% of total), and have the biggest heavy hitter of the trace (over 5% of the whole traffic).

The second set of 50 most active clients reveals a diffrent picture. Here, over 23 clients use only Web and Streaming, while the group of P2P users is much smaller. In these users, the *usual browsing* activity is much present with Web related clusters regrouping 2/3 of users.

It is interesting to see that P2P and Streaming users form very distinct groups as only 10 of 100 most active users mix these 2 applications. This is also the case with Download whose profile never overlaps P2P. This shows that there is a set of clients that prefer classical P2P and another set of clients that use one click hosting to download contents.

The clustering process indeed partitions the first 50 heavy hitters according to P2P first, whereas for the second 50 heavy hitters the first partition occurs on Web.

Application mix - discussion. Focusing on the first heavy hitters we observe that this family of users is dominated by P2P related heavy-hitters. Even if streaming activity can also lead a user to become a heavy user, the main part of the volume generated by this class comes from a majority of moderate streaming users.

We conjecture that this situation will persist as the popularity of streaming continues to increase. Indeed, this increase of popularity is likely to translate into more users streaming more videos rather than a few users streaming a lot. If the main content providers switch to High Definition video encoding (which has bit-rates up to 4 times larger than standard definition), this could have a dramatic impact for ISPs.

5 Conclusion

In this paper, we have proposed and illustrated several simple techniques to profile residential customers, with respect to their application level characteristics.

We have first presented an approach where the focus is on the dominant application of a user, which is justified by the fact that the dominant application explains a large majority of bytes for most users (in our data sets at least). This approach enables us to observe overall trends among moderately heavy and heavy users in a platform. We have next focused more deeply on the heavy hitters. Those heavy hitters are mostly P2P users, even though the global trend of traffic shows that Web and Streaming classes dominate. It is however understandable as P2P applications naturally tend to generate a few heavy hitters, while Web and Streaming tend to increase the volume of traffic of the average user.

We also devised an approach that seeks for common application mixes among the most active users of the platform. To this aim, we defined per application thresholds to differentiate real usage of an application from measurement artifacts. We use hierarchical clustering, that groups customers into a limited number of usage profiles. By focusing on the 100 most active users, divided in two equal sets, we demonstrated that:

- P2P users (pure P2P or mixed with other applications) are dominant in number and volume among the first 50 most active users;
- whereas in the second set of 50 most active users, the killer application is the combination of Web and Streaming.

Moreover while almost all P2P bytes are generated by the first 50 most active users, the Web + Streaming class is used by many users, and generates a fraction of bytes comparable (or higher) to P2P.

Our study sheds light on the traffic profile of the most active users in a residential platform, which has many implications for ISPs. However, we have only scratched the surface of the problem. Application at a larger scale of similar techniques, e.g., on much longer traces, would bring more insights than the snapshots we analyzed. As part of our future work, we plan to further extend the analysis, by tracking the evolution of users profiles on the long term.

The techniques presented in this paper complement the standard monitoring tools of the ISP platform and can help in predicting new trends in application usage. For instance, it gives us the root cause of faster and faster growth of Web + Streaming against P2P (usage democratization).

We strongly believe that hierarchical clustering on discretized attributes is a good approach because it greatly eases interpretation of the resulting clusters. Still, we plan to extend the discretization process from binary to (at least) ternary variables to take into account low/medium usage of an application *vs.* high usage.

References

1. Antoniades, D., Markatos, E.P., Dovrolis, C.: One-click hosting services: a file-sharing hideout. In: IMC (2009)
2. Bernaille, L., Teixeira, R., Salamatian, K.: Early application identification. In: CoNEXT (2006)

3. Breslau, L., Cao, P., Fan, L., Phillips, G., Shenker, S.: Web caching and Zipf-like distributions: evidence and implications. In: INFOCOM (1999)
4. Cho, K.: Broadband Traffic Report. Internet Infrastructure Review 4, 18–23 (2009)
5. Cho, K., Fukuda, K., Esaki, H., Kato, A.: The impact and implications of the growth in residential user-to-user traffic. In: SIGCOMM (2006)
6. Erman, J., Gerber, A., Hajiaghayi, M.T., Pei, D., Spatscheck, O.: Network-aware forward caching. In: WWW (2009)
7. Feldmann, A., Greenberg, A., Lund, C., Reingold, N., Rexford, J., True, F.: Deriving traffic demands for operational IP networks: methodology and experience. IEEE/ACM Trans. Netw. 9(3), 265–280 (2001)
8. Fessant, F., Lemaire, V., Clrot, F.: Combining Several SOM Approaches in Data Mining: Application to ADSL Customer Behaviours Analysis. In: Data Analysis, Machine Learning and Applications (2008)
9. Han, J.: Data Mining: Concepts and Techniques, 2nd edn. Morgan Kaufmann Publishers Inc., San Francisco (2006)
10. Karagiannis, T., Papagiannaki, K., Taft, N., Faloutsos, M.: Profiling the end host. In: Uhlig, S., Papagiannaki, K., Bonaventure, O. (eds.) PAM 2007. LNCS, vol. 4427, pp. 186–196. Springer, Heidelberg (2007)
11. Maier, G., Feldmann, A., Paxson, V., Allman, M.: On dominant characteristics of residential broadband internet traffic. In: IMC (2009)
12. Pietrzyk, M., Costeux, J.-L., Urvoy-Keller, G., En-Najjary, T.: Challenging statistical classification for operational usage: the ADSL case. In: IMC (2009)
13. Pietrzyk, M., Urvoy-Keller, G., Costeux, J.-L.: Revealing the unknown ADSL traffic using statistical methods. In: COST (2009)
14. Plissonneau, L., En-Najjary, T., Urvoy-Keller, G.: Revisiting web traffic from a DSL provider perspective: the case of YouTube. In: ITC Specialist Seminar on Network Usage and Traffic (2008)
15. Siekkinen, M., Collange, D., Urvoy-Keller, G., Biersack, E.W.: Performance Limitations of ADSL Users: A Case Study. In: Uhlig, S., Papagiannaki, K., Bonaventure, O. (eds.) PAM 2007. LNCS, vol. 4427, pp. 145–154. Springer, Heidelberg (2007)
16. Tanimoto, T.: An elementary mathematical theory of classification and prediction. In: IBM Program IBCLF (1959)
17. Vu-Brugier, G.: Analysis of the impact of early fiber access deployment on residential Internet traffic. In: ITC 21 (2009)

Sub-Space Clustering and Evidence Accumulation for Unsupervised Network Anomaly Detection

Johan Mazel[1,2], Pedro Casas[1,2], and Philippe Owezarski[1,2]

[1] CNRS; LAAS; 7 avenue du colonel Roche,
F-31077 Toulouse Cedex 4, France
[2] Universite de Toulouse; UPS, INSA, INP, ISAE; UT1, UTM, LAAS;
F-31077 Toulouse Cedex 4, France
{jmazel,pcasashe,owe}@laas.fr

Abstract. Network anomaly detection has been a hot research topic for many years. Most detection systems proposed so far employ a supervised strategy to accomplish the task, using either signature-based detection methods or supervised-learning techniques. However, both approaches present major limitations: the former fails to detect unknown anomalies, the latter requires training and labeled traffic, which is difficult and expensive to produce. Such limitations impose a serious bottleneck to the development of novel and applicable methods in the near future network scenario, characterized by emerging applications and new variants of network attacks. This work introduces and evaluates an unsupervised approach to detect and characterize network anomalies, without relying on signatures, statistical training, or labeled traffic. Unsupervised detection is accomplished by means of robust data-clustering techniques, combining Sub-Space Clustering and multiple Evidence Accumulation algorithms to blindly identify anomalous traffic flows. Unsupervised characterization is achieved by exploring inter-flows structure from multiple outlooks, building filtering rules to describe a detected anomaly. Detection and characterization performance of the unsupervised approach is extensively evaluated with real traffic from two different data-sets: the public MAWI traffic repository, and the METROSEC project data-set. Obtained results show the viability of unsupervised network anomaly detection and characterization, an ambitious goal so far unmet.

Keywords: Unsupervised Anomaly Detection & Characterization, Clustering, Clusters Isolation, Outliers Detection, Filtering Rules.

1 Introduction

Network anomaly detection has become a vital component of any network in today's Internet. Ranging from non-malicious unexpected events such as flash-crowds and failures, to network attacks such as denials-of-service and worms spreading, network traffic anomalies can have serious detrimental effects on the

J. Domingo-Pascual, Y. Shavitt, and S. Uhlig (Eds.): TMA 2011, LNCS 6613, pp. 15–28, 2011.
© Springer-Verlag Berlin Heidelberg 2011

performance and integrity of the network. The principal challenge in automatically detecting and characterizing traffic anomalies is that these are a moving target. It is difficult to precisely and permanently define the set of possible anomalies that may arise, especially in the case of network attacks, because new attacks as well a new variants to already known attacks are continuously emerging. A general anomaly detection system should therefore be able to detect a wide range of anomalies with diverse structure, using the least amount of previous knowledge and information, ideally no information at all.

The problem of network anomaly detection has been extensively studied during the last decade. Two different approaches are by far dominant in current research literature and commercial detection systems: signature-based detection and supervised-learning-based detection. Both approaches require some kind of guidance to work, hence they are generally referred to as supervised-detection approaches. Signature-based detection systems are highly effective to detect those anomalies which they are programmed to alert on. However, these systems cannot defend the network against new attacks, simply because they cannot recognize what they do not know. Furthermore, building new signatures is expensive, as it involves manual inspection by human experts. On the other hand, supervised-learning-based detection uses labeled traffic data to train a baseline model for normal-operation traffic, detecting anomalies as patterns that deviate from this model. Such methods can detect new kinds of anomalies and network attacks not seen before, because these will naturally deviate from the baseline. Nevertheless, supervised-learning requires training, which is time-consuming and depends on the availability of labeled traffic data-sets.

Apart from detection, operators need to analyze and characterize network anomalies to take accurate countermeasures. The characterization of an anomaly is a hard and time-consuming task. The analysis may become a particular bottleneck when new anomalies are detected, because the network operator has to manually dig into many traffic descriptors to understand its nature. Even expert operators can be quickly overwhelmed if further information is not provided to prioritize the time spent in the analysis.

Contrary to current supervised approaches, we develop in this work a completely unsupervised method to detect and characterize network anomalies, without relying on signatures, training, or labeled traffic of any kind. The proposed approach permits to detect both well-known as well as completely unknown anomalies, and to automatically produce easy-to-interpret signatures that characterize them. The algorithm runs in three consecutive stages. Firstly, traffic is captured in consecutive time slots of fixed length ΔT and aggregated in IP flows (standard *5-tuples*). IP flows are additionally aggregated at different *flow-resolution* levels, using {IPaddress/netmask} as aggregation key. Aggregation is done either for IPsrc or IPdst. To detect an anomalous time slot, time-series Z_t are constructed for simple traffic metrics such as number of bytes, packets, and IP flows per time slot, using the different flow-resolutions. Any generic change-detection algorithm $\mathcal{F}(.)$ based on time-series analysis [1–5] is then applied to Z_t: at each new time slot, $\mathcal{F}(.)$ analyses the different time-series associated with each aggregation level, going from coarser to finer-grained resolution. Time slot t_0 is flagged as anomalous if $\mathcal{F}(Z_{t_0})$ triggers an alarm for any of the traffic metrics at any flow-resolution. Tracking anomalies from multiple metrics and at multiple aggregation levels (i.e. /8, /16, /24, /32 netmask) provides additional reliability to the change-detection algorithm, and permits to detect both single source-destination and distributed anomalies of very different characteristics.

The unsupervised detection and characterization algorithm begins in the second stage, using as input the set of flows in the flagged time slot. Our method uses robust and efficient clustering techniques based on Sub-Space Clustering (SSC) [8] and multiple Evidence Accumulation (EA) [9] to blindly extract the suspicious traffic flows that compose the anomaly. As we shall see, the simultaneous use of SSC and EA improves the power of discrimination to properly detect traffic anomalies. In the third stage of the algorithm, the evidence of traffic structure provided by the SSC and EA algorithms is further used to produce filtering rules that characterize the detected anomaly, which are ultimately combined into a new anomaly signature. This signature provides a simple and easy-to-interpret description of the problem, easing network operator tasks.

The remainder of the paper is organized as follows. Section 2 presents a very brief state of the art in the supervised and unsupervised anomaly detection fields, additionally describing our main contributions. Section 3 introduces the core of the proposal, presenting an in-depth description of the clustering techniques and detection algorithms that we use. Section 4 presents the automatic anomaly characterization algorithm, which builds easy-to-interpret signatures for the detected anomalies. Section 5 presents an exhaustive validation of our proposals, discovering and characterizing single source/destination and distributed network attacks in real network traffic from two different data-sets: the public MAWI traffic repository of the WIDE project [17], and the METROSEC project data-set [18]. Finally, section 6 concludes this paper.

2   Related Work and Contributions

The problem of network anomaly detection has been extensively studied during the last decade. Traditional approaches analyze statistical variations of traffic volume metrics (e.g., number of bytes, packets, or flows) and/or other specific traffic features (e.g. distribution of IP addresses and ports), using either single-link measurements or network-wide data. A non-exhaustive list of methods includes the use of signal processing techniques (e.g., ARIMA, wavelets) on single-link traffic measurements [1, 2], PCA [7] and Kalman filters [4] for network-wide anomaly detection, and Sketches applied to IP-flows [3, 6].

Our proposal falls within the unsupervised anomaly detection domain. Most work has been devoted to the Intrusion Detection field, focused on the well known KDD'99 data-set. The vast majority of the unsupervised detection schemes proposed in the literature are based on clustering and outliers detection, being [11–13] some relevant examples. In [11], authors use a single-linkage hierarchical clustering method to cluster data from the KDD'99 data-set, based on the standard Euclidean distance for inter-pattern similarity. Clusters are considered as normal-operation activity, and patterns lying outside a cluster are flagged as anomalies. Based on the same ideas, [12] reports improved results in the same data-set, using three different clustering algorithms: Fixed-Width clustering, an optimized version of k-NN, and one class SVM. Finally, [13] presents a combined density-grid-based clustering algorithm to improve computational complexity, obtaining similar detection results.

Our unsupervised algorithm presents several advantages w.r.t. current state of the art. First and most important, it works in a completely unsupervised fashion, which means that it can be directly plugged-in to any monitoring system and start to detect anomalies from scratch, without any kind of calibration. Secondly, it performs anomaly detection based not only on outliers detection, but also by identifying small-size clusters. This is achieved by exploring different levels of traffic aggregation, both at the source and destination of the traffic, which additionally permits to discover low-intensity and distributed attacks. Thirdly, it avoids the lack of robustness of general clustering techniques used in current unsupervised anomaly detection algorithms; in particular, it is immune to general clustering problems such as sensitivity to initialization, specification of number of clusters, or structure-masking by irrelevant features. Fourthly, the algorithm performs clustering in low-dimensional spaces, using simple traffic descriptors such as number of source IP addresses or fraction of SYN packets. This simplifies the analysis and characterization of the detected anomalies, and avoids well-known problems of sparse spaces when working with high-dimensional data. Finally, the method combines the multiple evidence of an anomaly detected in different sub-spaces to produce an easy-to-interpret traffic signature that characterizes the problem. This permits to reduce the time spent by the network operator to understand the nature of the detected anomaly.

3   Unsupervised Anomaly Detection

The unsupervised anomaly detection stage takes as input all the flows in the time slot flagged as anomalous, aggregated according to one of the different levels used in the first stage. An anomaly will generally be detected in different aggregation levels, and there are many ways to select a particular aggregation to use in the unsupervised stage; for the sake of simplicity, we shall skip this issue, and use any of the aggregation levels in which the anomaly was detected. Without loss of generality, let $\mathbf{Y} = \{\mathbf{y}_1, .., \mathbf{y}_n\}$ be the set of n flows in the flagged time slot, referred to as *patterns* in more general terms. Each flow $\mathbf{y}_i \in \mathbf{Y}$ is described by a set of m traffic attributes or *features*. Let $\mathbf{x}_i = (x_i(1), .., x_i(m)) \in \mathbb{R}^m$ be the corresponding vector of traffic features describing flow $\mathbf{y}_i$, and $\mathbf{X} = \{\mathbf{x}_1, .., \mathbf{x}_n\}$ the complete matrix of features, referred to as the *feature space*.

The algorithm is based on clustering techniques applied to $\mathbf{X}$. The objective of clustering is to partition a set of unlabeled patterns into homogeneous groups of similar characteristics, based on some measure of similarity. Our particular goal is to identify and to isolate the different flows that compose the anomaly flagged in the first stage. Unfortunately, even if hundreds of clustering algorithms exist [10], it is very difficult to find a single one that can handle all types of cluster shapes and sizes. Different clustering algorithms produce different partitions of data, and even the same clustering algorithm provides different results when using different initializations and/or different algorithm parameters. This is in fact one of the major drawbacks in current cluster analysis techniques: the lack of robustness.

To avoid such a limitation, we have developed a *divide & conquer* clustering approach, using the notions of clustering ensemble [15] and multiple clusterings combination. The idea is novel and appealing: why not taking advantage of the information provided by multiple partitions of $\mathbf{X}$ to improve clustering robustness and detection results? A clustering ensemble $\mathbf{P}$ consists of a set of multiple partitions P_i produced for the same data. Each of these partitions provides a different and independent evidence of data structure, which can be combined to construct a new measure of similarity that better reflects natural groupings. There are different ways to produce a clustering ensemble. We use Sub-Space Clustering (SSC) [8] to produce multiple data partitions, applying the same clustering algorithm to N different sub-spaces $\mathbf{X}_i \subset \mathbf{X}$ of the original space.

3.1 Clustering Ensemble and Sub-Space Clustering

Each of the N sub-spaces $\mathbf{X}_i \subset \mathbf{X}$ is obtained by selecting k features from the complete set of m attributes. To deeply explore the complete feature space, the number of sub-spaces N that are analyzed corresponds to the number of k-combinations-obtained-from-m. Each partition P_i is obtained by applying DB-SCAN [16] to sub-space $\mathbf{X}_i$. DBSCAN is a powerful density-based clustering algorithm that discovers clusters of arbitrary shapes and sizes [10], and it is probably one of the most common clustering algorithms along with the widely known k-means. DBSCAN perfectly fits our unsupervised traffic analysis, because it is not necessary to specify a-priori difficult to set parameters such as the number of clusters to identify. The clustering result provided by DBSCAN is twofold: a set of p clusters $\{C_1, C_2, .., C_p\}$ and a set of q outliers $\{o_1, o_2, .., o_q\}$. To set the number of dimensions k of each sub-space, we take a very useful property of monotonicity in clustering sets, known as the downward closure property: "if a collection of points is a cluster in a k-dimensional space, then it is also part of a cluster in any $(k - 1)$ projections of this space". This directly implies that, if there exists any evidence of density in $\mathbf{X}$, it will certainly be present in its lowest-dimensional sub-spaces. Using small values for k provides several advantages: firstly, doing clustering in low-dimensional spaces is more efficient and faster than clustering in bigger dimensions. Secondly, density-based clustering algorithms such as DBSCAN provide better results in low-dimensional spaces [10], because high-dimensional spaces are usually sparse, making it difficult to distinguish between high and low density regions. Finally, results provided by low-dimensional clustering are more easy to visualize, which improves the interpretation of results by the network operator. We shall therefore use $k = 2$ in our SSC algorithm, which gives $N = m(m - 1)/2$ partitions.

3.2   Combining Multiple Partitions Using Evidence Accumulation

Having produced the N partitions, the question now is how to use the information provided by the obtained clusters and outliers to isolate anomalies from normal-operation traffic. An interesting answer is provided in [9], where authors introduced the idea of multiple-clusterings Evidence Accumulation (EA). EA uses the clustering results of multiple partitions P_i to produce a new inter-patterns similarity measure which better reflects natural groupings. The algorithm follows a split-combine-merge approach to discover the underlying structure of data. In the **split** step, the N partitions P_i are generated, which in our case they correspond to the SSC results. In the **combine** step, a new measure of similarity between patterns is produced, using a *weighting* mechanism to combine the multiple clustering results. The underlying assumption in EA is that patterns belonging to a "natural" cluster are likely to be co-located in the same cluster in different partitions. Taking the membership of pairs of patterns to the same cluster as weights for their association, the N partitions are mapped into a $n \times n$ similarity matrix S, such that $S(i,j) = n_{ij}/N$. The value n_{ij} corresponds to the number of times that pair $\{\mathbf{x}_i, \mathbf{x}_j\}$ was assigned to the same cluster in the N partitions. Note that if a pair of patterns $\{\mathbf{x}_i, \mathbf{x}_j\}$ is assigned to the same cluster in each of the N partitions then $S(i,j) = 1$, which corresponds to maximum similarity.

We adapt the EA algorithm for our particular problem of unsupervised anomaly detection. For doing so, let us think about the particular structure of any general anomaly. By simple definition of what it is, an anomaly may consist of either outliers or small-size clusters, depending on the aggregation level of flows in $\mathbf{Y}$. Let us take a flooding attack as an example; in the case of a DoS, all the packets of the attack will be aggregated into a single flow $\mathbf{y}_i$ targeting the victim, which will be represented as an outlier in $\mathbf{X}$. If we now consider a DDoS launched from β attackers towards a single victim, then the anomaly will be represented as a cluster of β flows if the aggregation is done for IPsrc/32, or as an outlier if the aggregation is done for IPdst/32. Taking into account that the number of flows in $\mathbf{Y}$ can reach a couple of thousands even for small time slots, the value of β would have to be too large to violate the assumption of small-size cluster.

We have developed two different EA methods to isolate small-size clusters and outliers: EA for small-clusters identification, EA4C, and EA for outliers identification, EA4O. Algorithm 1 presents the pseudo-code for both methods. In EA4C, we assign a stronger similarity weight when patterns are assigned to small-size clusters. The weighting function $w_k(n_t(k))$ used to update $S(i,j)$ at each iteration t takes bigger values for small values of $n_t(k)$, and goes to zero for big values of $n_t(k)$, being $n_t(k)$ the number of flows inside the co-assigned cluster for pair $\{\mathbf{x}_i, \mathbf{x}_j\}$. Parameters $n_{\min}$ and ρ specify the minimum number of flows that can be classified as a cluster and the neighborhood-density distance used by DBSCAN respectively. The parameter γ permits to set the slope of $w_k(n_t(k))$. Even tunable, we shall work with fixed values for $n_{\min}$, ρ, and γ, $n_{\min} = 20$, $\rho = 0.1$, and $\gamma = 5$, all empirically obtained.

Algorithm 1. EA4C & EA4O for Unsupervised Anomaly Detection

1: **Initialization:**
2:     Set similarity matrix S to a null $n \times n$ matrix.
3:     Set dissimilarity vector D to a null $n \times 1$ vector.
4: **for** $t = 1 : N$ **do**
5:     $P_t = \text{DBSCAN}(\mathbf{X}_t, n_{\min,\rho})$
6:     Update $S(i,j)$, $\forall$ pair $\{\mathbf{x}_i, \mathbf{x}_j\} \in C_k$ and $\forall C_k \in P_t$:
7:     $w_k \leftarrow e^{-\gamma \dfrac{(n_t(k) - n_{\min})}{n}}$
8:     $S(i,j) \leftarrow S(i,j) + \frac{w_k}{N}$
9:     Update $D(i)$, $\forall$ outlier $o_i \in P_t$:
10:     $w_t \leftarrow \dfrac{n}{(n - n_{\max_t}) + \epsilon}$
11:     $D(i) \leftarrow D(i) + \text{d}_{\text{M}}(o_i, C_{\max_t})\, w_t$
12: **end for**

In the case of EA4O, we define a dissimilarity vector D where the distances from all the different outliers to the centroid of the biggest cluster identified in each partition P_t are accumulated. We shall use $C_{\max_t}$ as a reference to this cluster. The idea is to clearly highlight those outliers that are far from the normal-operation traffic in the different partitions, statistically represented by $C_{\max_t}$. The weighting factor w_t takes bigger values when the size $n_{\max_t}$ of $C_{\max_t}$ is closer to the total number of patterns n, meaning that outliers are more rare and become more important as a consequence. The parameter ϵ is simply introduced to avoid numerical errors ($\epsilon = 1e^{-3}$). Finally, instead of using a simple Euclidean distance, we compute the Mahalanobis distance $\text{d}_{\text{M}}(o_i, C_{\max_t})$ between the outlier and the centroid of $C_{\max_t}$, which is an independent-of-features-scaling measure of similarity.

In the final **merge** step, any clustering algorithm can be applied to matrix S or to vector D to obtain a final partition of $\mathbf{X}$ that isolates both small-size clusters and outliers. As we are only interested in finding the smallest-size clusters and the most dissimilar outliers, the detection consists in finding the flows with the biggest similarity in S and the biggest dissimilarity in D. This is simply achieved by comparing the values in S and D to a variable detection threshold.

4   Automatic Characterization of Anomalies

At this stage, the algorithm has identified a set of traffic flows in $\mathbf{Y}$ far out the majority of the traffic. The following task is to automatically produce a set of K filtering rules $f_k(\mathbf{Y})$, $k = 1, .., K$ to correctly isolate and characterize these flows. In the one hand, such filtering rules provide useful insights on the nature of the anomaly, easing the analysis task of the network operator. On the other hand, different rules can be combined to construct a signature of the anomaly, which can be used to detect its occurrence in the future, using a traditional signature-based detection system. Even more, this signature could eventually be compared against well-known signatures to automatically classify the anomaly.

In order to produce filtering rules $f_k(\mathbf{Y})$, the algorithm selects those subspaces $\mathbf{X}_i$ where the separation between the anomalous flows and the rest of the traffic is the biggest. We define two different classes of filtering rule: *absolute* rules $f_A(\mathbf{Y})$ and *relative* rules $f_R(\mathbf{Y})$. Absolute rules are only used in the characterization of small-size clusters. These rules do not depend on the separation between flows, and correspond to the presence of dominant features in the flows of the anomalous cluster. An absolute rule for a certain feature j has the form $f_A(\mathbf{Y}) = \{\mathbf{y}_i \in \mathbf{Y} : x_i(j) == \lambda\}$. For example, in the case of an ICMP flooding attack, the vast majority of the associated flows use only ICMP packets, hence the absolute filtering rule {nICMP/nPkts == 1} makes sense.

On the contrary, relative filtering rules depend on the relative separation between anomalous and normal-operation flows. Basically, if the anomalous flows are well separated from the rest of the clusters in a certain partition P_i, then the features of the corresponding sub-space $\mathbf{X}_i$ are good candidates to define a relative filtering rule. A relative rule defined for feature j has the form $f_R(\mathbf{Y}) = \{\mathbf{y}_i \in \mathbf{Y} : x_i(j) < \lambda \text{ or } x_i(j) > \lambda\}$. We shall also define a *covering relation* between filtering rules: we say that rule f_1 *covers* rule $f_2 \leftrightarrow f_2(\mathbf{Y}) \subset f_1(\mathbf{Y})$. If two or more rules overlap (i.e., they are associated to the same feature), the algorithm keeps the one that covers the rest.

In order to construct a compact signature of the anomaly, we have to devise a procedure to select the most discriminant filtering rules. Absolute rules are important, because they define inherent characteristics of the anomaly. As regards relatives rules, their relevance is directly tied to the degree of separation between flows. In the case of outliers, we select the K features for which the Mahalanobis distance to the normal-operation traffic is among the top-K biggest distances. In the case of small-size clusters, we rank the degree of separation to the rest of the clusters using the well-known Fisher Score (FS), and select the top-K ranked rules. The FS measures the separation between clusters, relative to the total variance within each cluster. Given two clusters C_1 and C_2, the Fisher Score for feature i can be computed as:

$$F(i) = \frac{(\bar{x}_1(i) - \bar{x}_2(i))^2}{\sigma_1(i)^2 + \sigma_2(i)^2}$$

where $\bar{x}_j(i)$ and $\sigma_j(i)^2$ are the mean and variance of feature i in cluster C_j. In order to select the top-K relative rules, we take the K features i with biggest $F(i)$ value. To finally construct the signature, the absolute rules and the top-K relative rules are combined into a single inclusive predicate, using the covering relation in case of overlapping rules.

5   Experimental Evaluation in Real Traffic

We evaluate the ability of the unsupervised algorithm to detect and to construct a signature for different attacks in real traffic from the public MAWI repository of the WIDE project [17]. The WIDE operational network provides interconnection between different research institutions in Japan, as well as connection to different commercial ISPs and universities in the U.S.. The traces we shall work with consist of 15 minutes-long raw packet traces collected at one of the trans-pacific links between Japan and the U.S.. Traces are not labeled, thus our analysis is limited to show the detection and characterization of different network attacks found by manual inspection in randomly selected traces, such as ICMP DoSs, SYN network scans, and SYN DDoS. Whenever possible, we refer to results obtained in [6], where some of these attacks have already been identified.

We shall also test the true positive and false positive rates obtained with annotated attacks, using different traffic traces from the METROSEC project [18]. These traces consist of real traffic collected on the French RENATER network, containing simulated attacks performed with well-known DDoS attack tools. Traces were collected between 2004 and 2006, and contain DDoS attacks that range from very low intensity (i.e., less than 4% of the overall traffic volume) to massive attacks (i.e., more than 80% of the overall traffic volume). Additionally, we shall compare the performance of the algorithm against some traditional methods for unsupervised outliers detection presented in section 2, and also against the very well-known PCA and the sub-space approach [7].

In these evaluations we use the following list of $m = 9$ traffic features: number of source/destination IP addresses and ports (nSrcs, nDsts, nSrcPorts, nDstPorts), ratio of number of sources to number of destinations, packet rate (nPkts/sec), fraction of ICMP and SYN packets (nICMP/nPkts, nSYN/nPkts), and ratio of packets to number of destinations. According to previous work on signature-based anomaly characterization [14], such simple traffic descriptors permit to describe standard attacks such as DDoS, scans, and spreading worms. The list is by no means exhaustive, and more features can be easily plugged-in to improve results. In fact, a paramount advantage of our approach is that it is not tied to any particular set of features, and can therefore be generalized to any kind of traffic descriptors. For $m = 9$ features, we get $N = 36$ sub-spaces to analyze, a pretty small clustering ensemble that can be computed very fast.

5.1   Detecting Attacks in the Wild: MAWI Traffic

We begin by detecting and characterizing a distributed SYN network scan directed to many victim hosts under the same /16 destination network. The trace consists of traffic captured the 01/04/01. Traffic in **Y** is aggregated in IPdst/24 flows, thus we shall detect the attack as a small-size cluster. To appreciate the great advantage of using the SSC-EA-based algorithm w.r.t. a traditional approach, based on directly clustering the complete feature space, we shall compute a "traditional" similarity matrix S_{tra} for the n flows in **Y**. Each element $S_{\text{tra}}(i,j)$ represents inter-flows similarity by means of the Euclidean distance. Figures 1.(a,b) depict the discrimination provided by both similarity matrices S and S_{tra}, using a Multi-Dimensional Scaling (MDS) analysis. The anomalous flows are mixed-up with the normal ones w.r.t. S_{tra}, and the discrimination using all the features at the same time becomes difficult. In the case of S, the flows

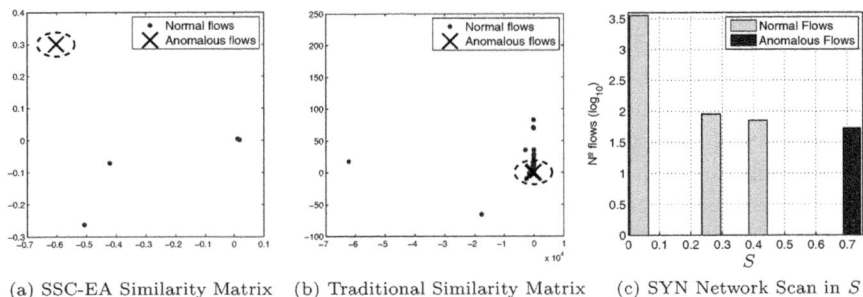

(a) SSC-EA Similarity Matrix (b) Traditional Similarity Matrix (c) SYN Network Scan in S

Fig. 1. MDS for traditional and SSC-EA-based clustering. A SYN network scan can be easily detected using the SSC-EA similarity measure.

that compose the attack are perfectly isolated from the rest, providing a powerful discrimination. As we explained before, the detection of the attack consists in identifying the most similar flows in S. Figure 1.(c) depicts a histogram on the distribution of inter-flows similarity, according to S. Selecting the most similar flows results in a compact cluster of 53 flows. A further analysis of the traffic that compose each of these flows reveals different IPdst/32 sub-flows of SYN packets with the same origin IP address, corresponding to the attacker.

Regarding filtering rules and the characterization of the attack, figures 2.(a,b) depict some of the partitions P_i where both absolute and relative filtering rules where found, corresponding to those with biggest Fisher Score. These rules involve the number of IP sources and destinations, and the fraction of SYN packets. Combining them produces a signature that can be expressed as (nSrcs == 1) ∧ (nDsts > λ_1) ∧ (nSYN/nPkts > λ_2), where λ_1 and λ_2 are two thresholds obtained by separating the clusters at half distance. This signature makes perfect sense, since the network scan uses SYN packets from a single attacking host to a large number victims. The signature permits to correctly identify all the flows of the attack. The beauty and main advantage of the unsupervised approach relies on the fact that this new signature has been produced without any previous information about the attack or the baseline traffic.

The next two case-studies correspond to flooding attacks. For practical issues, traffic corresponds to different combined traces (14/10/03, 13/04/04, and 23/05/06). Figures 2.(c,d) depict different rules obtained in the detection of a SYN DDoS attack. Traffic is now aggregated in IPsrc/32 flows. The distribution analysis of inter-flows similarity w.r.t. S selects a compact cluster with the most similar flows, corresponding to traffic from the set of attacking hosts. The obtained signature can be expressed as (nDsts == 1) ∧ (nSYN/nPkts > λ_3) ∧ (nPkts/sec > λ_4), which combined with the large number of identified sources (nSrcs > λ_5) confirms the nature of a SYN DDoS attack. This signature is able to correctly isolate the most aggressive hosts of the DDoS attack, namely those with highest packet rate. Figures 2.(e,f) depict the detection of an ICMP flooding DoS attack. Traffic is aggregated using aggregation index IPdst/32, thus the attack is detected as an outlier rather than as a small cluster. Besides showing

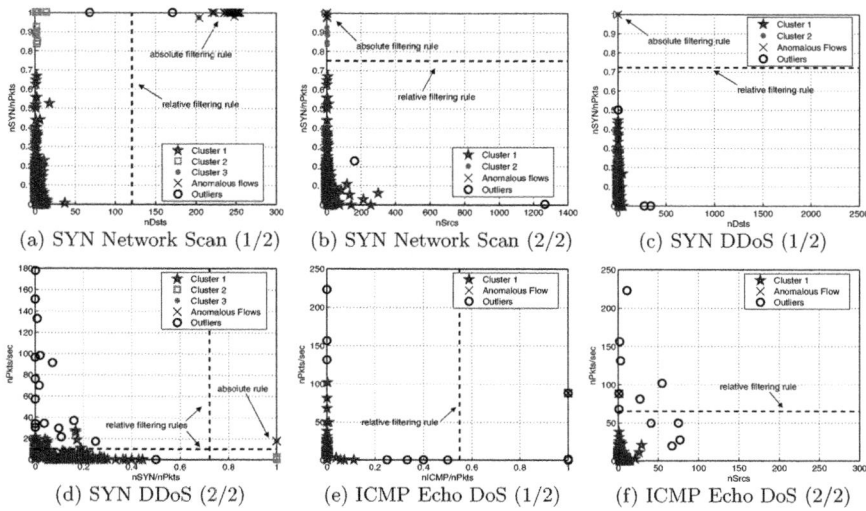

Fig. 2. Filtering rules for characterization of attacks in MAWI

typical characteristics of this attack, such as a high packet rate of exclusively ICMP packets from the same source host, both partitions show that the detected attack does not represent the largest elephant flow in the time slot. This emphasizes the ability of the algorithm to detect low volume attacks, even of lower intensity than normal traffic. The obtained signature can be expressed as $(nICMP/nPkts > \lambda_6) \wedge (nPkts/sec > \lambda_7)$.

To conclude, we present the detection of two different attacks in one of the traces previously analyzed in [6], captured the 18/03/03. Traffic is aggregated in IPsrc/32 flows, and both attacks are detected as outliers. Figure 3.(a) shows the ordered dissimilarity values in D obtained by the EA4O method, along with their corresponding label. The first two most distant flows correspond to a highly distributed SYN network scan (more than 500 destination hosts) and an ICMP spoofed flooding attack directed to a small number of victims (ICMP redirect packets towards port 0). The following two flows correspond to unusual large rates of DNS traffic and HTTP requests; from there on, flows correspond to normal-operation traffic. The ICMP flooding attack and the two unusual flows are also detected in [6], but the SYN scan was missed by their method. Note that both attacks can be easily detected and isolated from the anomalous but yet legitimate traffic without false alarms, using for example the threshold α_1 on D. Figures 3.(b,c) depict the corresponding four flows in two of the N partitions produced by EA4O. As before, we verify the ability of the algorithm to detect network attacks that are not necessary the biggest elephant flows.

5.2   Detecting Attacks with Ground Truth: METROSEC Traffic

Figure 4 depicts the True Positives Rate (TPR) vs. the False Positives Rates (FTR) in the detection of 9 DDoS attacks in the METROSEC data-set.

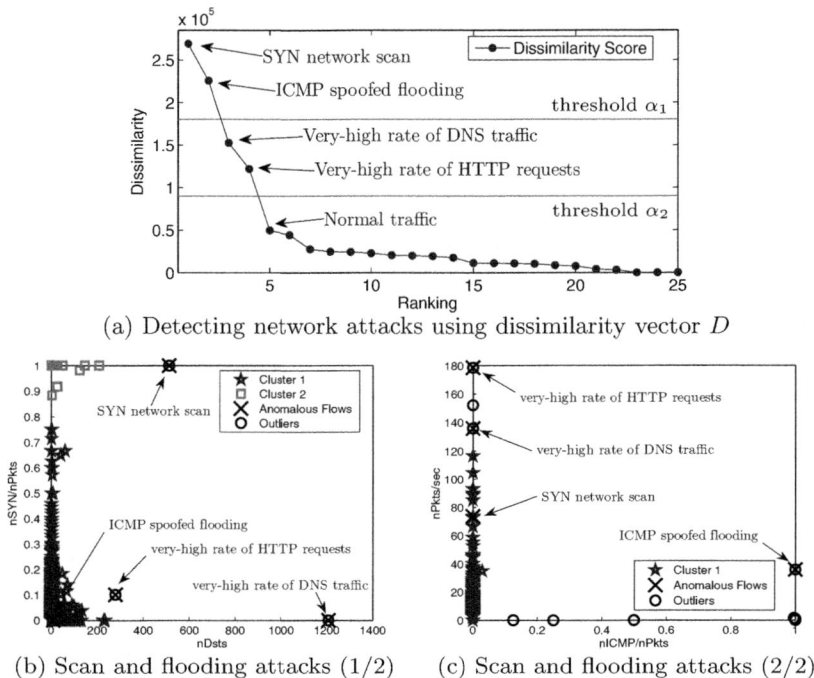

(a) Detecting network attacks using dissimilarity vector D

(b) Scan and flooding attacks (1/2)   (c) Scan and flooding attacks (2/2)

Fig. 3. Detection and analysis of network attacks in MAWI

Detection is performed with traffic aggregated at the IPdst/32 level. Traffic corresponds to different combined traces (24/11/04, 07-09-10/12/04, and 11/04/06). The ROC plot is obtained by comparing the sorted dissimilarities in D to a variable detection threshold. From the 9 documented attacks, 5 correspond to massive attacks (more than 70% of traffic), 1 to a high intensity attack (about 40%), 2 are low intensity attacks (about 10%), and 1 is a very-low intensity attack (about 4%). EA4O correctly detects 8 out of the 9 attacks without false alarms. The detection of the very-low intensity attack is more difficult; however, the 9 attacks are correctly detected with a very low FPR, about 1.2%.

We compare the performance of our approach against three "traditional" approaches: DBSCAN-based, k-means-based, and PCA-based outliers detection. The first two consist in applying either DBSCAN or k-means to the complete feature space $\mathbf{X}$, identify the largest cluster C_{max}, and compute the Mahalanobis distance of all the flows lying outside C_{max} to its centroid. The ROC is finally obtained by comparing the sorted distances to a variable detection threshold. These approaches are similar to those used in previous work [11–13]. In the PCA-based approach, PCA and the sub-space methods [7] are applied to the complete matrix $\mathbf{X}$, and the attacks are detected by comparing the residuals to a variable threshold. Both the k-means and the PCA-based approaches require fine tuning: in k-means, we repeat the clustering for different values of clusters k, and take the average results. In the case of PCA we present the best performance, obtained for 2 principal components to describe the normal sub-space.

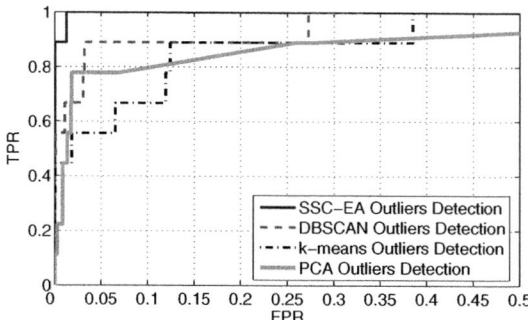

Fig. 4. DDoS detection in METROSEC. EA4O detects even low intensity attacks with a very-low false alarm rate, which is not achieved by traditional approaches.

Obtained results permit to evidence the great advantage of using the SSC-EA-based algorithm in the clustering step w.r.t. to traditional approaches. In particular, all the approaches used in the comparison fail to detect the smallest attack with a reasonable false alarm rate. Both clustering algorithms based either on DBSCAN or k-means get confused by masking features when analyzing the complete feature space **X**. The PCA approach shows to be not sensitive enough to discriminate both low-intensity and high-intensity attacks, using the same representation for normal traffic.

6   Concluding Remarks

The completely unsupervised anomaly detection algorithm that we have presented has many interesting advantages w.r.t. previous proposals in the field. It uses exclusively unlabeled data to detect and characterize network anomalies, without assuming any kind of signature, particular model, or canonical data distribution. This allows to detect new previously unseen anomalies, even without using statistical-learning. Despite using ordinary clustering techniques, the algorithm avoids the lack of robustness of general clustering approaches, by combining the notions of Sub-Space Clustering and multiple Evidence Accumulation. The Sub-Space Clustering approach also permits to obtain easy-to-interpret results, providing insights and explanations about the detected anomalies to the network operator. Even more, clustering in low-dimensional feature spaces provides results that can be visualized by standard techniques, which improves the assimilation of results.

We have verified the effectiveness of our proposal to detect and isolate real single source/destination and distributed network attacks in real traffic traces from different networks, all in a completely blind fashion, without assuming any particular traffic model, significant clustering parameters, or even clusters structure beyond a basic definition of what an anomaly is. Additionally, we have shown detection results that outperform traditional approaches for outliers detection, providing a stronger evidence of the accuracy of the SSC-EA-based method to detect network anomalies.

Acknowledgments

This work has been done in the framework of the ECODE project, funded by the European commission under grant FP7-ICT-2007-2/223936.

References

1. Barford, P., et al.: A Signal Analysis of Network Traffic Anomalies. In: Proc. ACM IMW (2002)
2. Brutlag, J.: Aberrant Behavior Detection in Time Series for Network Monitoring. In: Proc. 14th Systems Administration Conference (2000)
3. Krishnamurthy, B., et al.: Sketch-based Change Detection: Methods, Evaluation, and Applications. In: Proc. ACM IMC (2003)
4. Soule, A., et al.: Combining Filtering and Statistical Methods for Anomaly Detection. In: Proc. ACM IMC (2005)
5. Cormode, G., et al.: What's New: Finding Significant Differences in Network Data Streams. IEEE Trans. on Networking 13(6), 1219–1232 (2005)
6. Dewaele, G., et al.: Extracting Hidden Anomalies using Sketch and non Gaussian Multiresolution Statistical Detection Procedures. In: Proc. SIGCOMM LSAD (2007)
7. Lakhina, A., et al.: Diagnosing Network-Wide Traffic Anomalies. In: Proc. ACM SIGCOMM (2004)
8. Parsons, L., et al.: Subspace Clustering for High Dimensional Data: a Review. ACM SIGKDD Expl. Newsletter 6(1), 90–105 (2004)
9. Fred, A., et al.: Combining Multiple Clusterings Using Evidence Accumulation. IEEE Trans. Pattern Anal. and Machine Intel. 27(6), 835–850 (2005)
10. Jain, A.K.: Data Clustering: 50 Years Beyond K-Means. Pattern Recognition Letters 31(8), 651–666 (2010)
11. Portnoy, L., et al.: Intrusion Detection with Unlabeled Data Using Clustering. In: Proc. ACM DMSA Workshop (2001)
12. Eskin, E., et al.: A Geometric Framework for Unsupervised Anomaly Detection: Detecting Intrusions in Unlabeled Data. In: Apps. of Data Mining in Comp. Sec., Kluwer Publisher, Dordrecht (2002)
13. Leung, K., et al.: Unsupervised Anomaly Detection in Network Intrusion Detection Using Clustering. In: Proc. ACSC 2005 (2005)
14. Fernandes, G., et al.: Automated Classification of Network Traffic Anomalies. In: Proc. SecureComm 2009 (2009)
15. Strehl, A., et al.: Cluster Ensembles - A Knowledge Reuse Framework For Combining Multiple Partitions. Jour. Mach. Learn. Res. 3, 583–617 (2002)
16. Ester, M., et al.: A Density-based Algorithm for Discovering Clusters in Large Spatial Databases with Noise. In: Proc. ACM SIGKDD (1996)
17. Cho, K., et al.: Data Repository at the WIDE Project. In: USENIX ATC (2000)
18. METROlogy for SECurity and QoS, http://laas.fr/METROSEC

An Analysis of Longitudinal TCP Passive Measurements (Short Paper)

Kensuke Fukuda

National Institute of Informatics,
Japan / PRESTO JST
kensuke@nii.ac.jp

Abstract. This paper reports on the longitudinal dynamics of TCP flows at an international backbone link over the period from 2001 to 2010. The dataset was a collection of traffic traces called MAWI data consisting of daily 15min pcap traffic trace measured at a trans-pacific link between Japan and the US. The environment of the measurement link has changed in several aspects (i.e., congestion, link upgrade, application). The main findings of the paper are as follows. (1) A comparison of the AS-level delays between 2001 and 2010 shows that the mean delay decreased in 55% of ASes, but the median value increased. Moreover, largely inefficient paths disappeared. (2) The deployment of TCP SACK increased from 10% to 90% over the course of 10 years. On the other hand, the window scale and timestamp options remained under-deployed (less than 50%).

1 Introduction

The Internet is one of the most important infrastructures in our daily life, and a better understanding of its status will be crucial for ISPs and end-users. ISPs need traffic characteristics for conducting daily operations and making decisions about the future deployments. End-users are sometimes annoyed with their low throughput or large delay. The stability of the network is affected by many issues; changes made to the access/backbone links, evolution of applications, changes in wide-area routing, etc. In general, it is hard to predict the future behavior of a network, so an investigation of the network traffic for a long period would be useful for gaining a better understanding of the network behavior.

Numerous studies have been conducted on many aspects of the network [7,12,2,6,4,8]. In this paper, we analyze longitudinal TCP traffic dynamics by using passive measurement data collected at a trans-pacific transit link between Japan and the US during the period from 2001 to 2010. In early part of the period, excessive traffic caused congestion on the link, but then the link upgrade in 2006 mitigated it. Also, a wide variety of the application protocols were observed. We believe that longitudinal traffic traces are enough for investigating the characteristics of the network dynamics. In this paper we especially focus on the time evolution of the delay, connection status, and usage of TCP options.

The main findings of the paper are as follows; (1) 60% of the established TCP connections were correctly closed, though 20% of the web connections were

J. Domingo-Pascual, Y. Shavitt, and S. Uhlig (Eds.): TMA 2011, LNCS 6613, pp. 29–36, 2011.
© Springer-Verlag Berlin Heidelberg 2011

terminated with reset with FIN flags. The same was observed in higher port traffic (i.e., p2p), however, its ratio decreased rapidly over time. (2) The delay between ASes was strongly affected by congestion on the link as expected. A comparison of the AS-level delays between 2001 and 2010 shows that the mean delay decreased in 55% of the ASes, but the median increased over the same period. Furthermore, we found that largely inefficient paths have disappeared. (3) The deployment of TCP SACK increased from 10% to 90% over 10 years. On the other hand, the window scale and timestamp options are still under the deployment (less than 50%).

2 Related Work

Passive and active measurements of network traffic have been widely used in order to clarify the many aspects of the network (i.e., delay, jitter, loss, throughput, available bandwidth, etc.).

For passive TCP-based measurements, there are many studies devoted to accurately estimate the RTT in three-way-hand shake phase or slow-start phase [10,9]. In terms of the connection status of TCP it has been shown that some web server and client pairs finish their connection with FIN and RST packets instead of normal termination with 2 FINs [3]. Similar observations have been reported in the behavior of certain P2P software [11]. Furthermore, Refs. [13,5,11] focused on the deployment of TCP options.

There have been a few studies focusing on the time evolution of traffic dynamics. Ref. [7] reported basic statistics of several backbone network traffic traces from 1998-2003. Furthermore, other studies have investigated time evolution of specific traffic type; anomalies [2,8], P2P [12,1], and self-similarity [4].

3 Dataset

The data set we analyzed was composed of MAWI traces, which are daily 15 min pcap traces (14:00-14:15) collected at a transit link including a trans-pacific

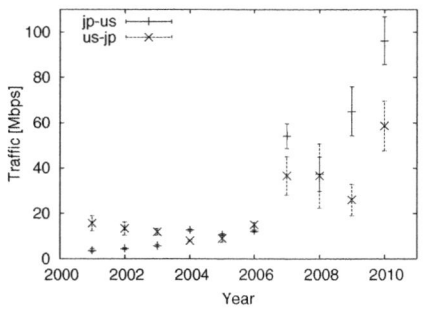

Fig. 1. Traffic volume

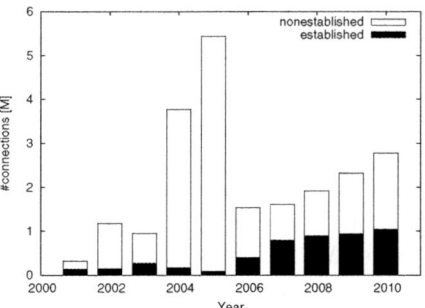

Fig. 2. Number of connections

link between Japan and the US during 2001-2010 with tcpdump command. Each trace consisted of pcap format packets without payload. We used seven consecutive original traces on the 1st week of March during 2001-2010. We believed that the data would be enough for understanding in the longitudinal behavior of Internet traffic, because certain aspects of the network traffic in MAWI traffic have been already analyzed, including anomalies [6,8] and long-range dependency [4]. The most dominant application in the traces is web throughout, and also clear and hidden P2P traffic is non-negligible. The application traffic breakdown is shown in MAWI web page (http://mawi.wide.ad.jp/mawi) and Ref. [6].

Figure 1 shows the average traffic of the link (a) from JP to US, and (b) from US to JP as identified by the MAC address of the packets in the traces. The error bars in the figure indicate the standard deviations of the traffic calculated from a time series of the traffic volume in 1s time bins. The link bandwidth was originally 18Mbps CAR, but then was upgraded to 100Mbps in July 2006. As pointed out in [4], the link was highly congested before the upgrade, showing small standard deviations. On the other hand, the traffic volume after the upgrade indicates high variability (large standard deviation). The ratio of corrpt packets at the link was high up to 1.2% of the packets during congested periods (2003-2006), but largely decreased ($\approx$ 0.0001%) after the upgrade.

4   Analysis of Longitudinal Traffic

4.1   Status of Connections

We shall first examine the connection status of TCP over time. TCP establishes a connection with 2 SYN packets (three-way hand shake; 3whs) and closes the connection with 2 FIN packets. Tracking TCP flags enables us to categorize the connection behavior into typical patterns. Figure 2 displays the number of established and non-established TCP connections. An established connection is defined as the connection with 2 SYN packets, and a non-established connection fails to establish the connection in the 3whs phase. The number of established connections increased from 138K in 2001 to 1M in 2010. A large number of non-established connections in 2004 and 2005 were due to scanning activities by viruses/worms. Specifically, the dominant TCP ports were 135, 445, 80, and 3127 (mainly mydoom) in 2004, and 9898, 5554, 135, and 445 (mainly sasser) in 2005.

Close Behavior of Established Connections. We shall deal with the following types of TCP closing behavior for established connections [3,11]:

- 2FIN: a connection normally closed with 2 FIN packets.
- RST: a connection was reset by either of the hosts.
- RST+FIN: one sent a FIN for closing, but the other replied a RST. This can be observed in a pair of web server and browser [3], and P2P nodes [11].
- 1FIN: one sent a FIN packet but there was no response from the other.
- unclosed: a connection lasts during a measurement time period.

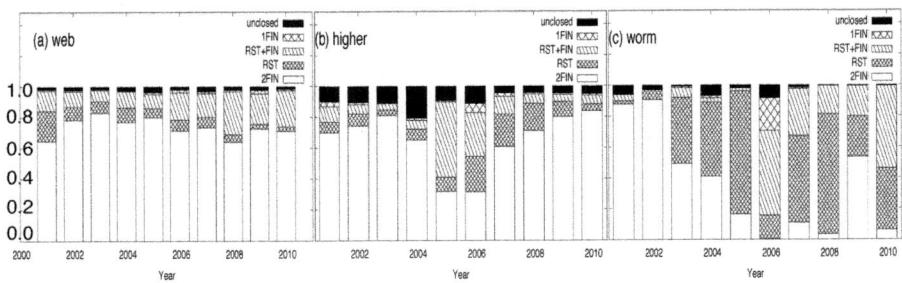

Fig. 3. Breakdown of established connections

Figure 3 is a breakdown of the closing behaviors of the established connections; (a) web connections ($port = 80, 443, 8080$), (b) connections with higher ports ($ports \geq 1024$), and (c) connections with ports known to be used by virus/worm. For web connections, complete connections (2FIN) accounted for 60-80% during the observed periods. The RST decreased gradually over time, and 10-20% of the TCP flows closed with RST+FIN which appeared in some pairs of Windows server-clients [11]. On the other hand, connections with higher ports show intrinsic behavior. The number of unclosed connections was large during 2001-2005. This is to be expected because P2P traffic flows are long lived. Also, the higher port connections included long-lived FTP-related traffic in earlier periods. A large portion of the RST+FIN flows was mainly due to the P2P software with port 6343. A similar observation was reported in [11], in which some P2P software saved resources by sending RST instead of FIN. In addition, the minor contribution of recent RST+FIN flows suggests that P2P software has shifted its behavior from RST+FIN to 2FIN. Moreover, we confirmed that one of the contributions to the RST behavior was related to spam assassin. Finally, the behavior of the worm flows (d) is clearly dissimilar to others. The flows with 2FIN were dominant in 2001-2002, and most of them were legitimate, though we could not distinguish between legitimate web and code red worm with port 80. However, from 2003, 20-80% of RST flows were with port 139, 445, and 1433. The increase in RST+FIN in 2006-2007 as well as 1FIN in 2006 was mainly due to Blaster worm (port 1433, 4444, and 4899). Thus, the worm types of flows were mainly characterized in terms of RST and RST+FIN.

Non-Established Connections. Non-established connections are classified into (a) only SYN packet and (b) SYN and corresponding RST packet. More than 90% of only SYN behavior were due to worms, 90% for higher ports, and 50-90% for Web. The high ratio of SYN and RST packets for web was due to scanning to the web server. The ratio was stable over the observed period.

4.2   Delay Behavior

Here, we investigate the delays as calculated from TCP flows. There are a number of algorithms to estimate the RTT. A simple and reliable estimation algorithm, called Syn-Ack (SA) method, is to use the delays in the 3whs phase [10]. In order to process the data with the SA method, we first identified the direction of the link from the MAC address of the packet, and then translated the source and destination IP addresses of the packet into AS numbers by extracting BGP routing information extracting from the routeview data. Consequently, we obtained the average round trip delay from the measurement point to each ASes.

Figure 4 shows the complementary cumulative distribution of delays over time in log-log scale. All curves are characterized by a long-tail and close to a power-law decay; most hosts communicate with relatively closer (in time) hosts (e.g., 90% of the RTTs are less than 500ms), though we still observed longer delays over 1s with a certain probability. The tails of the curves were longer during congested periods (2006) than during non-congested periods, although the link upgrade resulted in faster decaying tails.

In order to understand the average behavior of the delay, Fig. 5 presents the mean and median of the delay calculated per AS. The means of the delay are relatively stable; the mean in 2001 was close to that in 2010, and the effect of the congestion did not clearly appear in the mean delay. On the other hand, the median has increasing trend independent of the link upgrade. In particular, during 2006-2010, the median increased from 200ms to 300ms despite the decrease in the mean from 250ms to 180ms. Thus, although fewer extremely long delays (i.e., insufficient path) occurred, a large number of ASes were nonetheless characterized by a longer delay.

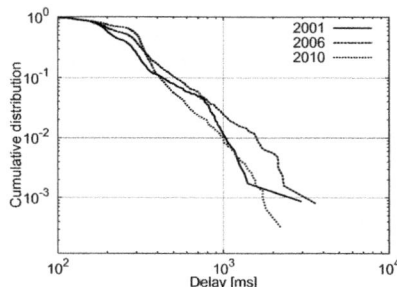 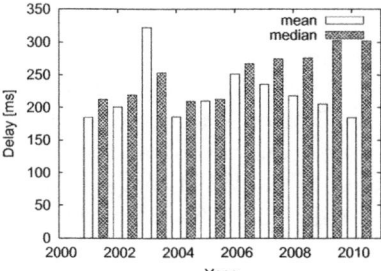

Fig. 4. Complementary cumulative distribution of estimated Delays

Fig. 5. Mean and median of Delays

Furthermore, to investigate ASes with longer delays, Fig. 6 displays the scatter plots of the mean delay between (a) 2001 and 2006, (b) 2006 and 2010, and (c) 2001 and 2010 for ASes commonly appearing in both datasets. The region below the diagonal line indicates the reduction in delay. The different symbols in the figure indicate the geographical difference of the ASes based on Internet registries. The results roughly reflect the geographical distance of the ASes (ARIN < RIPE < LACNIC < AFRNIC) from Japan. We also observed large changes in the delays; During 2001-2006, the delay of the ASes in LACNIC from Japan significantly decreased. Also, during 2006 and 2010, the delay in APNIC apparently decreased. In this sense, insufficient paths must have decreased in

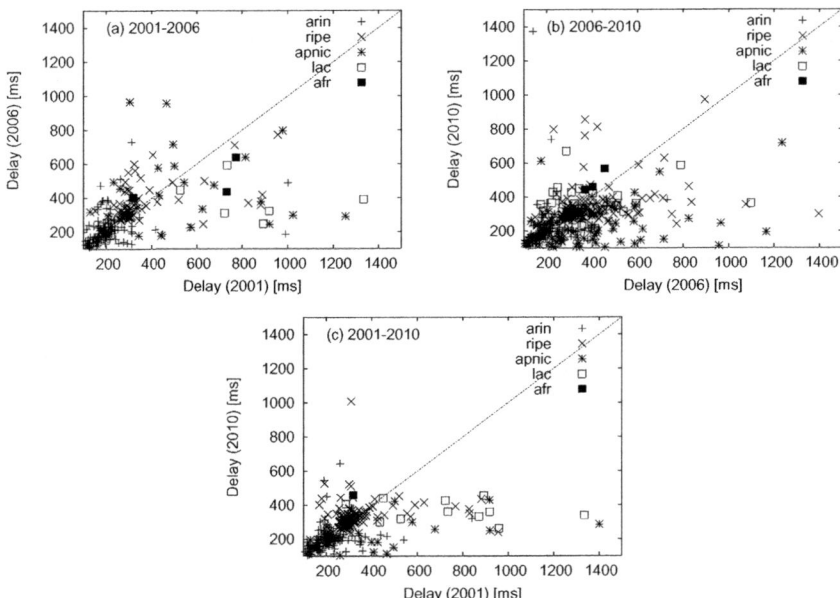

Fig. 6. Scatter plot of AS-level delay in different 2 years; (a) 2001 and 2006, (b) 2006 and 2010, and (c) 2001 and 2010

number during the observed period. However, the fact that we saw larger delays suggests that the Internet paths have a large variability and still there is likely more room to optimize the global routes. From 2001 to 2006, 55% of the ASes increased their delay, and only 60% of them decreased it from 2006 to 2010. As a whole, 55% of ASes decreased their delay from 2001 to 2010.

We also confirmed that the AS path length calculated from the BGP data collected by the WIDE project showed that the average shortest AS paths were stable (mean: 3.4-3.6hops, median: 3hops) in 2004-2010. Thus, the change in the RTTs cannot be explained only by the AS hops.

In summary, the largely inefficient paths have been less appeared over time, however, the median delay becomes larger from the observation of TCP flows.

4.3   TCP Options

Finally, we analyze the deployment and availability of TCP options that provide extensional functionalities for performance improvements. End hosts negotiate available TCP options in the 3whs phase; an initiator host sends a list of available TCP options: then a responder replies with the list of acceptable options. Consequently, the agreed-upon options are used in the connection. The options we analyzed were Max segment size (MSS), Selective ACK (SACK), Timestamp, and Window scale options. The possible results of the negotiation are (1) OK (i.e., both agreed), (2) NG (i.e., rejected by either of hosts), and (3) None (i.e, no option was exchanged).

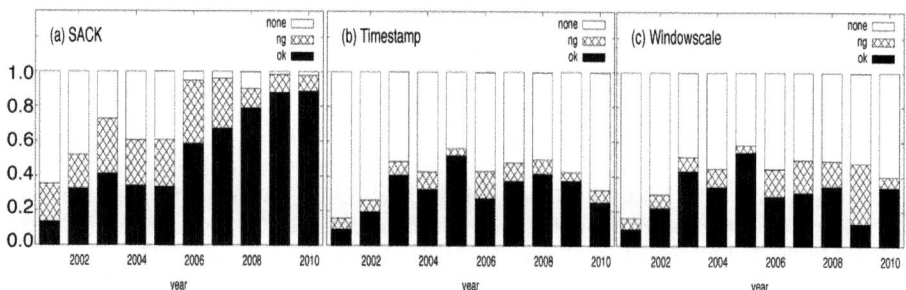

Fig. 7. Breakdown of TCP options

Figure 7 displays the status of TCP options for established connections. We omitted MSS options because it was fully deployed in 2001. The use of the SACK option has increased over time, and was available in 90% of the connections in 2010, compared with only 10% in 2001. Over 80% of SACK-denied hosts were web server (i.e., the source port is 80/8080/443). This finding is consistent with the previous study [11]; however, we found that the rejection ratio has recently decreased. A plausible reason of the high availability of the SACK option is its early deployment in Windows; it was turned on by default in Window 98.

On the other hand, the availability of timestamp and window scale options was less than 60% even in 2010. Window scale and timestamp options in Windows 2000/XP are turned off by default, but are enabled in recent Windows (Vista and 7). The options are enabled by default in Linux and BSD-inherent OSes. According to the breakdown of the OS type of the hosts in the data set by passive OS fingerprinting (p0f), we found that Windows 2000/XP hosts accounted for 77% of the total even in 2010, while recent Windows (Vista and Seven) hosts only accounted for 3%. The availability of those two options will increase in the near future as more of the recent Windows get deployed on hosts, and this eventuality will likely place higher throughputs on each TCP flows. However, we could not find a clear correlation between the value of the window scale and the throughput. Further investigations will be required before we can make a detailed quantification of this issue.

5   Conclusion

In this paper, we investigated the longitudinal characteristics of TCP flows from passive measurements made at a trans-pacific transit link over 10 years. The environment of the measurement link has changed in several aspects (i.e., congestion, link upgrade, applications). The main findings of the analysis are (1) the median RTT per ASes increased over time while the number of extremely insufficient paths are less appeared, and (2) the deployment and use of TCP SACK options has significantly increased, but time stamp and window scale options have yet to be fully utilized.

Acknowledgments. The author thanks Patrice Abry, Pierre Borgnat, Romain Fontugne, and Kenjiro Cho for their valuable comments.

References

1. Acosta, W., Chandra, S.: Trace driven analysis of the long term evolution of gnutella peer-to-peer traffic. In: Uhlig, S., Papagiannaki, K., Bonaventure, O. (eds.) PAM 2007. LNCS, vol. 4427, pp. 42–51. Springer, Heidelberg (2007)
2. Allman, M., Paxson, V., Terrell, J.: A brief history of scanning. In: IMC 2007, pp. 77–82 (2007)
3. Arlitt, M., Willamson, C.: An analysis of TCP reset behaviour on the internet. SIGCOMM CCR 35, 37–44 (2005)
4. Borgnat, P., Dewaele, G., Fukuda, K., Abry, P., Cho, K.: Seven years and one day: Sketching the evolution of internet traffic. In: IEEE INFCOM 2009, Rio de Janeiro, Brazil (2009)
5. Bykova, M., Ostermann, S.: Statistical analysis of malformed packets and their origins in the modern Internet. In: IMW 2002, pp. 83–88 (2002)
6. Dewaele, G., Fukuda, K., Borgnat, P., Abry, P., Cho, K.: Extracting hidden anomalies using sketch and non gaussian multiresolution statistical detection procedure. In: ACM SIGCOMM LSAD 2007, pp. 145–152 (2007)
7. Fomenkov, M., Keys, K., Moore, D., Claffy, K.: Longitudinal study of internet traffic in 1998-2003. In: WISICT 2004, Cancun, Mexico, pp. 1–6 (January 2004)
8. Fontugne, R., Borgnat, P., Abry, P., Fukuda, K.: Mawilab: Combining diverse anomaly detectors for automated anomaly labeling and performance benchmarking. In: ACM CoNEXT 2010, Philadelphia, PA, p. 12 (2010)
9. Jaiswal, S., Iannaccone, G., Diot, C., Kurose, J., Towsley, D.: Inferring TCP connection characteristics through passive measurements. In: INFOCOM 2004, Hong Kong, pp. 1582–1592 (March 2004)
10. Jiang, H., Dovrolis, C.: Passive estimation of TCP round-trip times. Computer Communications Review 32(3), 75–88 (2002)
11. John, W., Tafvelin, S., Olovsson, T.: Trends and differences in connection-behavior within classes of internet backbone traffic. In: Claypool, M., Uhlig, S. (eds.) PAM 2008. LNCS, vol. 4979, pp. 192–201. Springer, Heidelberg (2008)
12. Madhukar, A., Willamson, C.: A longitudinal study of p2p traffic classification. In: MASCOTS 2006, Monterey, CA, pp. 179–188 (September 2006)
13. Pentikousis, K., Badr, H.: Quantifying the deployment of TCP options – A comaprative study. IEEE Communications Letters 8(10), 647–649 (2004)

Understanding the Impact of the Access Technology: The Case of Web Search Services

Aymen Hafsaoui[1,*], Guillaume Urvoy-Keller[2], Denis Collange[3],
Matti Siekkinen[4], and Taoufik En-Najjary[3]

[1] Eurecom, Sophia-Antipolis, France
[2] Laboratoire I3S CNRS, Université de Nice, Sophia Antipolis, France
[3] Orange Labs, France
[4] Aalto University School of Science and Technology, Finland

Abstract. In this paper, we address the problem of comparing the performance perceived by end users when they use different technologies to access the Internet. We focus on three key technologies: Cellular, ADSL and FTTH. Users primarily interact with the network through the networking applications they use. We tackle the comparison task by focusing on Web search services, which are arguably a key service for end users. We first demonstrate that RTT and packet loss alone are not enough to fully understand the observed differences or similarities of performance between the different access technologies. We then present an approach based on a fine-grained profiling of the data time of transfers that sheds light on the interplay between service, access and usage, for the client and server side. We use a clustering approach to identify groups of connections experiencing similar performance over the different access technologies. This technique allows to attribute performance differences perceived by the client separately to the specific characteristics of the access technology, behavior of the server, and behavior of the client.

Keywords: TCP, Performances, Web search, User behaviors, Access Impact.

1 Introduction

Telecommunication operators offer several technologies to their clients for accessing the Internet. We have observed an increase in the offering of cellular and Fiber-To-The-Home (FTTH) accesses, which now compete with the older ADSL and cable modem technologies. However, until now it is unclear what are the exact implications of the significantly different properties of these access technologies on the quality of service observed by clients.

Our main objective in this paper is to devise a methodology to compare the performance of a given service over different access technologies. We consider three popular technologies to access the Internet: Cellular, ADSL, and FTTH. We use traces of end users traffic collected over these three types of access networks under the control of a major European ISP. We focus on an arguably key service for users: Web search engines, esp. Google and Yahoo.

---

* Corresponding author.

J. Domingo-Pascual, Y. Shavitt, and S. Uhlig (Eds.): TMA 2011, LNCS 6613, pp. 37–50, 2011.
© Springer-Verlag Berlin Heidelberg 2011

In this paper, we present a methodology to separately account for the impact of access, service usage, and application on top. The methodology is based on breaking down the duration of an entire Web transaction into sub-components which can be attributed to network or either of the end points. This kind of approach is vital because the typical performance metrics such as average latency, average throughput, and packet loss only give an overview of the performance but do not say much about what the origins are.

Our methodology can be applied in different ways depending on the objectives of the study. For example, a service provider might only want to analyze the performance contribution of the server, while an ISP could be more interested in the (access) network's contribution. In both cases, the focus of the study could be the performance observed by the majority of clients or, alternatively, troubleshooting through identification of performance anomalies. We exemplify various use cases for Yahoo and Google Web search services.

2   Related Work

Comparing the relative merits of different access technologies has been the subject of a number of studies recently. In [1], the authors analyze passive traffic measurements from ADSL and FTTH commercial networks under the control of the same ISP. They demonstrate that only a minority of clients and flows really take advantage of the high capacity of FTTH access. The main reason is the predominance of p2p protocols that do not exploit locality and high transmission capacities of other FTTH clients.

In [2], the authors investigate the benefits and optimizations of TCP splitting for accelerating cloud services, using web search as an exemplary case study and through an experimental system deployed in a production environment. They report that a typical response to an average search query takes between 0.6 and 1.0 second (between the TCP SYN and the last HTTP packet). The RTT between the client and the data-center during the measurement period was around 100 milliseconds. Search time within the data-center ranges almost uniformly between 50 and 400 msec[1]. Four TCP windows are required to transfer the result page to the client when there is no packet loss. The total time taken in this case is 5RTT + search time.

In [3], the authors present results from a measurement campaign for GPRS, EDGE, cellular, and HSDPA radio access, to evaluate the performance of web transfers with and without caching. Results were compared with the ones of a standard ADSL line (down:1Mb/s; up:256kb/s). Benchmarks reveal that there is a visible gain introduced by proxies within the technologies: HSDPA is often close to ADSL but does not outperform it; In EDGE, the proxy achieves the strongest improvement, bringing it close to HSDPA performance.

In [4], the authors quantify the improvement provided by a 3G access compared to 2G access in terms of delays and throughput. They demonstrate that for wired access networks (ADSL and FTTH) the average number of servers accessed per subscriber is one order of magnitude lower on the mobile trace, esp. because of the absence of P2P traffic. Focusing on the user experience when viewing multimedia content, they show how their behavior differs and how the radio access type influences their performance.

---

[1] We observe a significant fraction of values outside of this range in Section 6.

In [5] authors analyze Web search clickstreams by extracting the HTTP headers and bodies from packet-level traffic. They found that most queries consist of only one keyword and make little use of search operators, users issue on average four search queries per session, of which most consecutive ones are distinct. Relying on a developed Markov model that captures the logical relationships of the accessed Web pages authors reported additional insights on users' Web search behavior.

In [6] Stamou and all studied how web information seekers pick the search keywords to describe their information needs and specifically examine whether query keyword specifications are influenced by the results the users reviewed for a previous search. Then, they propose a model that tries to capture the results' influence on the specification of the subsequent user queries.

3   Data Sets

We study three packet level traces of end users traffic from a major French ISP involving different access technologies: ADSL, cellular and FTTH. ADSL and FTTH traces correspond to all the traffic of an ADSL and FTTH Point-of-Presence (PoP) respectively, while the cellular trace is collected at a GGSN level, which is the interface between the mobile network and the Internet. The cellular corresponds to 2G and 3G/3G+ accesses as clients with 3G/3G+ subscriptions can be downgraded to 2G depending on the base station capability. Table 1 summarizes the main characteristics of each trace.

Table 1. Traces Description

	cellular	FTTH	ADSL
Date	2008-11-22	2008-09-30	2008-02-04
Starting Capture	13:08:27	18:00:01	14:45:02:03
Duration	01:39:01	00:37:46	00:59:59
Number of Connections	1772683	574295	594169
Well-behaved connections	1236253	353715	381297
Volume Upload(GB)	11.2	51.3	4.4
Volume Download(GB)	50.6	74.9	16.4

In the present work, our focus is on applications on top of TCP, which carries the vast majority of bytes in our 3 traces, and close to 100% for the cellular technology. We restrict our attention to the connections that correspond to presumably valid and complete transfers, that we term well-behaved connections. Well-behaved connections must fulfill the following conditions: (i) A complete three-way handshake; (ii) At least one TCP data segment in each direction; (iii) The connection must finish either with a FIN or RESET flag. Well-behaved connections carry between 20 and 125 GB of traffic in our traces (see Table 1).

4   Web Search Traffic: A First Look

In this section, we focus on the traffic related to Google Web Search engine, which is the dominant Web Search engine in our traces. We focus here on the overall performance metrics before introducing our methodology for finer grained analysis in Section 5. We compare the Google and Yahoo cases in Section 6.

To identify traffic generated by the usage of Google search engine, we isolate connections that contain the string www.google.com/fr in their HTTP header. Relying simply on information at the IP and TCP layers would lead to incorporate in our data set other services offered by Google like gmail or Google map, which are serviced by the same IPs.

To identify Google search traffic for the upstream and downstream directions, we use TCP port numbers and remote address resolution. Table 2 summarizes the amount of Google search traffic we identified in our traces. We observed that FTTH includes the smallest number of such connections among the three traces, one explanation of this phenomenon was the short duration of the FTTH trace.

Table 2. Google Search Traffic in the Traces

	Cellular	FTTH	ADSL
Well-behaved Connections	29874	1183	6022
Data Packets Upload	107201	2436	18168
Data Packets Download	495374	7699	139129
Volume Upload(MB)	74.472	1.66	11.39
Volume Download(MB)	507.747	8	165.79

4.1 Connection Size

Figure 1(a) depicts the cumulative distribution of well-behaved Google search connection size in bytes. It appears that data transfer sizes are very similar for the three access technologies. This observation constitutes a good starting point since the performance of TCP depends on the actual transfer size. RTTs and losses also heavily influence TCP performance, as the various TCP throughput formulas indicate [7, 8]. Also, the available bandwidth plays a role. With respect to these metrics, we expect the performance of a service to be significantly influenced by the access technology since available bandwidth, RTTs[2] and losses are considerably different over ADSL, FTTH and Cellular. However, as we demonstrate in the remaining of this section, those metrics alone fail to fully explain the performance observed in our traces.

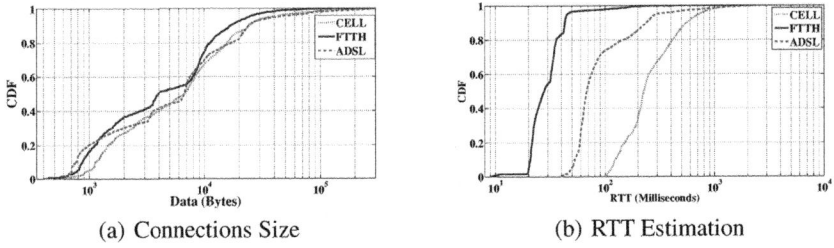

(a) Connections Size   (b) RTT Estimation

Fig. 1. General Performances

4.2 Latency

Several approaches have been proposed to accurately estimate the RTT from a single measurement point [11–15]. We considered two such techniques. The first method is

---

[2] As noted in several studies on ADSL [9] and Cellular networks[10], the access technology often contributes to a significant fraction of overall RTT.

based on the observation of the TCP 3-way handshake [12]: one first computes the time interval between the SYN and the SYN-ACK segment, and adds to the latter the time interval between the SYN-ACK and its corresponding ACK. The second method is similar but applied to TCP data and acknowledgement segments transferred in each direction[3]. One then takes the minimum over all samples as an estimate of the RTT. It is important to note that we take losses into account in our analysis (see next section).

We observed that both estimation methods (SYN-/SYN-ACK and DATA-ACK) lead to the same estimates except for the case of cellular access because of a Performance Enhancing Proxy (PEP) which biased the results from the SYN-/SYN-ACK method, as the PEP responds to SYN packets from the clients on behalf of the servers. We thus rely on the DATA-ACK method to estimate RTTs over the 3 technologies. Figure 1(b) depicts the resulting RTT estimations for the 3 traces (for Google Web search service only). It clearly highlights the impact of the access technology on the RTT. FTTH access offer very low RTT in general – less than 50 ms for more than 96% of connections. This finding is in line with the characteristics generally advertised for FTTH access technology. In contrast, RTTs on the Cellular technology are notably longer than under ADSL and FTTH.

4.3   Packet Loss

To assess the impact of TCP loss retransmission times on the performance of Google Web search traffic, we developed an algorithm to detect retransmitted data packets, which happen between the capture point and the server or between the capture point and the client. This algorithm[4] is similar to the one developed in [11].

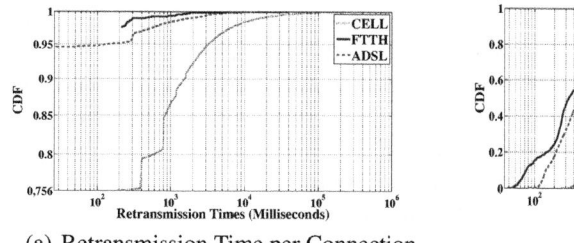
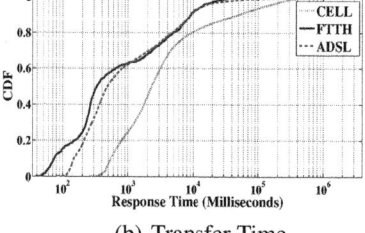

(a) Retransmission Time per Connection   (b) Transfer Time

Fig. 2. Immediate Access Impacts

If ever the loss happens after the observation point, we observed the initial packet and its retransmission. In this case, the retransmission time is simply the duration between

---

[3] Keep in mind that we focus on well-behaved transfers for which there is at least one data packet in each direction. Hence, we can apply the second method.

[4] The used loss' detection algorithm is available on http://intrabase.eurecom.fr/tmp/papers.html. People are invited to check the correctness of our algorithm to detect losses.

those two epochs[5]. When the packet is lost before the probe, we infer the epoch at which it should have been observed, based on the sequence numbers of packets. We try to separate real retransmission from network out of sequence events by eliminating durations smaller than the RTT of the connection.

Figure 2(a) depicts the cumulative distribution of retransmission time per connection for each trace. Retransmissions are clearly more frequent for the cellular access with more than 25% of transfers experiencing losses compared to less than 6% for ADSL and FTTH accesses. From previous works, we noticed that several factors explain high loss ratio for cellular access. In fact, in [16] authors recommend to use a loss detection algorithm, which uses dumps of each peer of the connection (this algorithm is not adapted for our case because our measurements have been collected at a GGSN level) to avoid spurious Retransmission Timeouts in TCP. In addition, authors report in [10] that spurious retransmission ratio, for SYN and ACK packets, in cellular networks is more higher for Google servers than other ones, due to short implemented Timeouts.

Most of the transfers are very short in terms of number of packets and we know that for such transfers, packet loss has a detrimental impact to the performance[17]. Thus, the performance of these transfers are dominated by the packet loss. In Sections 5.3 and 6, we analyze all connections, including the ones that experience losses by first removing recovery times from their total duration.

4.4   Application Level Performance

Our study of the two key factors that influence the throughput of TCP transfers , namely loss rate and RTT, suggest that, since Google Web search transfers have a similar profile on the 3 access technologies, the performance of this service over FTTH should significantly outperform the one of ADSL, which should in turn outperform the one of Cellular. It turns out that reality is slightly more complex as can be seen from Figure 2(b) where we report the distribution of transfer times (the figure for throughput is qualitatively similar but we prefer to report transfer times since Web search is an interactive service). Indeed, while the Cellular technology offers significantly longer response time, in line with RTT and loss factors, FTTH and ADSL have much closer performance than RTT and loss were suggesting.

In the next section, we propose a new analysis method that uncovers the impact of specific factors like the application and the interaction with user, and thus informs the comparison of access technology.

5   Interplay between Application, Usage and Access

The analysis method that we use consists in two steps. In the first step, the transfer time of each TCP connection is broken down into several factors that we can attribute to different causes, e.g., the application or the end-to-end path. In a second step, we use a clustering approach to uncover the major trends within the different data sets under study.

---

[5] Those epochs are computed at the sender side by shifting the time series according to our RTT estimate.

5.1   Step 1: Data Time Break-Down

For this first step, we introduce a methodology that has been initially proposed in [17]. The objective is to reveal the impact of each layer that contributes to the overall data transfer time, namely the application, the transport, and the end-to-end path (network layer and layers below) between the client and the server.

The starting point is that the vast majority of transfers consist of dialogues between the two sides of a connection, where each party talks in turn. This means that application instances rarely talk simultaneously on the same TCP connection [17]. We call the sentences of these dialogues *trains*.

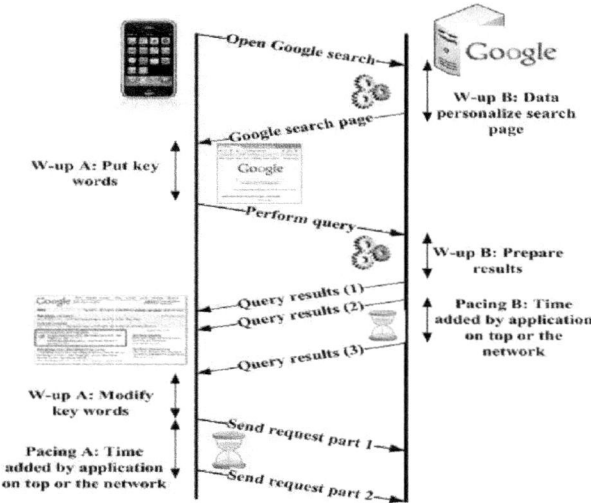

Fig. 3. Data Time Break-Down

We term A and B the two parties involved in the transfer (A is the initiator of the transfer) and we break down the data transfer into three components: warm-up time, theoretical time and pacing time. Figure 3 illustrates this break down in the case of a Google search where A is a client of the ISP and B is a Google server.

A warm-up corresponds to the time taken by A or B before answering to the other party. It includes durations such as thinking time at the user side or data preparation at the server side. For our use case, a warm-up of A corresponds to the time spent by the client to type a query and to browse through the results before issuing the next query (if any) or clicking on a link, whereas a warm-up of B corresponds to the time spent by the Google server to prepare the appropriate answer to the request.

Theoretical time is the duration that an ideal TCP transfer would take to transfer an amount of packets from A to B (or from B to A) equal to the total amount of packets exchanged during the complete transfer. Theoretical time can be seen as the total transfer time of this ideal TCP connection that would have all the data available right at the beginning of the transfer. For this ideal transfer, we further assume that the capacity of the path is infinite and an RTT equal to RTT_{A-B} (or RTT_{B-A}).

Once warm-up and theoretical times have been substracted from the total transfer time, some additional time may remain. We term that remaining time pacing time. While theoretical time can be attributed to characteristics of the path and warm-up time to applications and/or user, pacing factors effects due either to the access link or some mechanism higher up in the protocol stack. Indeed, as we assume in the computation of theoretical time that A and B have infinite access bandwidth, we in fact assume that we can pack as many MSS size packets within an RTT as needed, which is not necessarily true due to a limited access bandwidth. In this case, the extra time will be factored in the pacing time. Similarly, if the application or some middle-boxes are throttling the transmission rate, this will also be included in the pacing time. A contextual interpretation that accounts for the access and application characteristics is thus needed to uncover the cause behind observed pacing time. The above breakdown of total transfer time is computed for each side A and B separately.

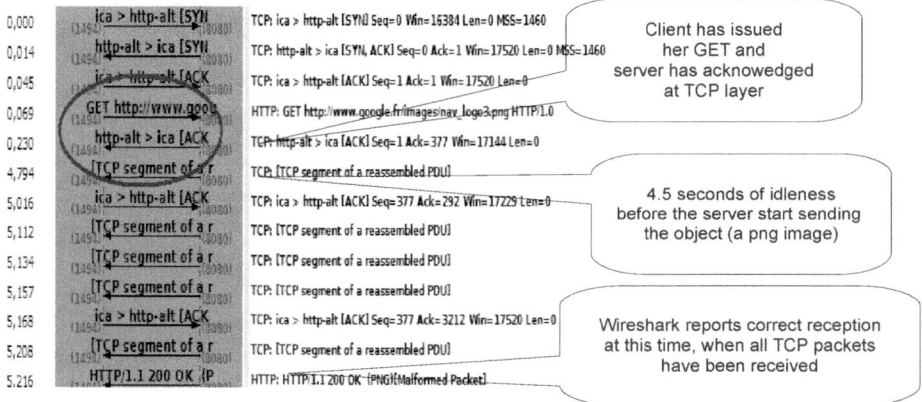

Fig. 4. Abnormal Long Response Time at The Server Side (Warm-up B value)

We report on Figure 4 an example of observed large warm-up time at the server side, for a client behind an ADSL access. We noticed that the acknowledgement received from the server indicates that the query (GET request) has been correctly received by the server, but it takes about 4.5 seconds before the client starts to receive the requested object (a png image in this case). As we can see next in Section 5.3, an easy identification of these extreme cases can be a useful application of our methodology.

5.2   Step 2: Data Clustering

The second analysis step is new as compared to our previous work [17]. For this second step, we use clustering approaches to obtain a global picture of the relation between the service, the access technology and the usage.

At the end of step 1, each well-behaved Google search connection is transformed into a point in a 6-dimensional space (pacing, theoretical and train time of the client and the server). To mine this data, we use a clustering technique to group connections with similar characteristics. We use an unsupervised clustering approach as we have no a priori knowledge of the characteristics of the data to be analyzed, e.g., a model of normal and abnormal traffic. We chose the popular Kmeans algorithm. A key issue when using Kmeans is the choice of the initial centroids and the number of clusters targeted. Concerning the choice of the centroids, we perform one hundred trials and take the best result, i.e., the one that minimizes the sum over all clusters of the distances between each point and its centroid.

To assess the number of clusters, we rely on a visual dimensionality reduction technique, t-SNE (t-Distributed Stochastic Neighbour Embedding)[18]. t-SNE projects multi-dimensional data on a plane while preserving the inner neighbouring characteristics of data. Application of t-SNE to our 6-dimensional data leads to the right plot of Figure 5(a). This figure indicates that a natural clustering exists within our data. In addition, a reasonable value for the number of clusters lies between 5 and 10. Last but not least the right plot of Figure 5(a) suggests that some clusters are dominated by a specific access technology while some others are mixed. We picked a value of 6 for the number of clusters in Kmeans. Note that we use the matlab implementation of Kmeans [19].

5.3   Results

Figure 5(b) depicts the 6 clusters obtained by application of Kmeans. We use boxplots[6] to obtain compact representations of the values corresponding to each dimension. We indicate, on top of each cluster, the number of samples in the cluster for each access technology. We use the same number of samples per access technology to prevent any bias in the clustering, which limits us to 1000 samples, due to the short duration of the FTTH trace. The ADSL and Cellular samples were chosen randomly among the ones in the respective traces. We also plot in Figure 6(b) the size of the transfers of each cluster and their throughput[7].

We first observe that the clusters obtained with Kmeans are in good agreement with the projection obtained by t-SNE as indicated in the left plot of Figure 5(a), where data samples are indexed using their cluster id in Kmeans.

Before delving into the interpretation of the individual clusters, we observe that three of them carry the majority of the bytes. Indeed, Figure 6(a) indicates that clusters 1 and 2 and 6 represent 83% of the bytes. Let us first focus on these dominant clusters.

Clusters 1, 2 and 6 are characterized by large warm-up A values, i.e., long waiting time at the client side in between two consecutive requests. The warm-up A values are in the order of a few seconds, which are compatible with human actions. This behavior is in line with the typical use of search engines where the user first submits a query then analyzes the results before refining further her query or clicking on one of the links of the result page. Thus, the primary factor that influences observed throughputs in Google search traffic is the user behavior. In fact, identified values in clusters 1, 2 and 6 of Warm-up A are in line with results in [6] of the time between query submission and first click, where authors identified different users trends.

---

[6] Boxplots are compact representations of distributions: the central line is the median and the upper and lower of the box the 25th and 75th quantiles. Extreme values -far from the waist of the distribution - are reported as crosses.

[7] We compute the throughput by excluding the tear down time, which is the time between the last data packet and the last packet of the connection. This specific metric that we term Application Level (AL) throughput offers a more accurate view of the user experience [17].

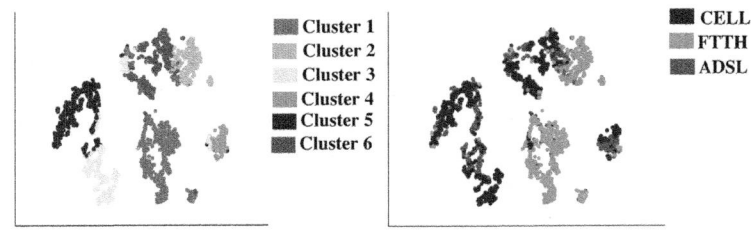

(a) T-SNE

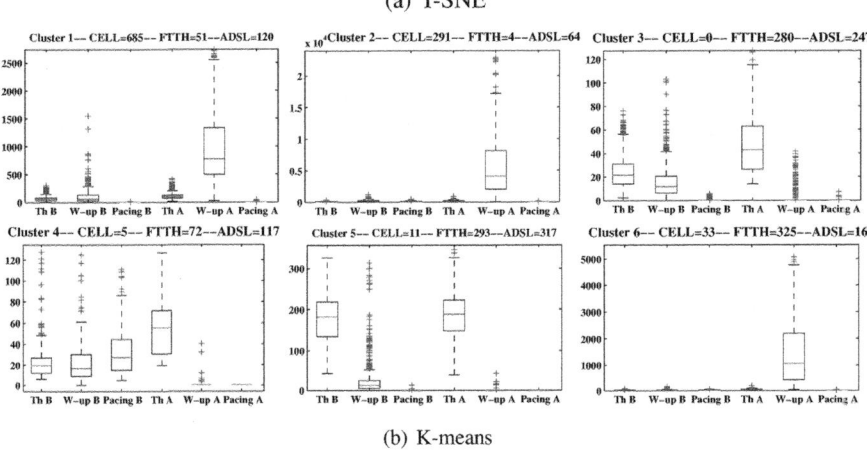

(b) K-means

Fig. 5. Google Search Engine Clusters

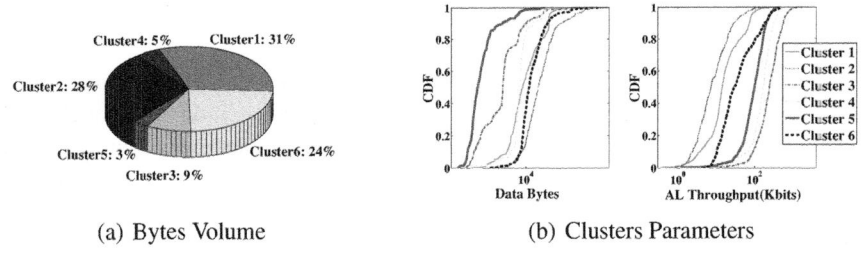

(a) Bytes Volume   (b) Clusters Parameters

Fig. 6. Google Search Engine Parameters

We can further observe that clusters 1 and 2 mostly consist of cellular connections while cluster 6 consists mostly of FTTH transfers. This means that the clustering algorithm first based its decision on the Warm-up A value; then, this is the access technology that impacts the clustering. As ADSL offers intermediate characteristics as compared to FTTH and Cellular, ADSL transfers with large Warm-up A values are scattered on the three clusters.

Let us now consider clusters 3, 4 and 5. Those clusters, while carrying a tiny fraction of traffic, feature several noticeable characteristics. First, we see almost no cellular connections in those clusters. Second, they total two thirds of the ADSL and FTTH connections, even though they are smaller than the ones in clusters 1, 2 and 6 – see Figure 6(b). Third, those clusters, in contrast to clusters 1, 2 and 6 have negligible Warm-up A values. From a technical viewpoint, Kmeans separates them based on the RTT as cluster 5 exhibits larger ThA and ThB values and also based on Pacing B values. After a further analysis of these clusters we observed that they corresponds to very short connection with an exchange of 2 HTTP frames, Google servers finish current connection after an idle period of 10 seconds. Moreover, cluster 3 presents cases when client opens Google web search page in their Internet browser without performing any search request, then after a time-out of 10 seconds Google server close the connection. In other hand, cluster 4 and 5 corresponds to Get request and HTTP OK response with an effective search, the main difference between cluster 4 and 5 were RTT and connection size.

More generally, we expect that our method, when applied to profile other services, will lead to some clusters that can be easily related to the behavior of the service under study while some others will relate anomalous or unsual behaviors that might require further investigation. For the case of Google search engine, we do not believe cluster 3,4,5 are anomalies per se that affects the quality of experience of users since the large number of connections in those clusters would prevent the problem from flying below the radar. We found only very few cases where the server's impact to the performance was dominating and directly impacting the quality of experience of the end user. Observing many such cases would have indicated issues, e.g., with service implementation or provisioning.

6   Contrasting Web Search Engines

The main idea in this section is to contrast Google results with others Web search services. For the case of our traces, we observed that the second dominant Web Search engine is Yahoo, though with an order of magnitude less connections. This low number of samples somehow limits the applicability of our clustering approach as used in the Google case. We restrict our attention to the following questions: (i) Do the two services offer similar traffic profile? (ii) Are services provisioned in a similar manner? Architecture of Google and Yahoo data-centers are obviously different but they must both obey the constraint that the client must receive its answer to a query in a maximum amount of time that is in the order of a few hundreds of milliseconds [2]. We investigate the service provisioning by analyzing the Warm-up B values (data preparation time at server) offered by the two services.

6.1   Traffic Profiles

Figure 7(a) shows cdfs of data connections size for Cellular, FTTH and ADSL traces for both Google and Yahoo. We observe for our traces that Yahoo Web search connections are larger than Google ones. An intuitive explanation behind this observation is that Yahoo Web search pages contain, on average, more photos and banners than ordinary Google pages.

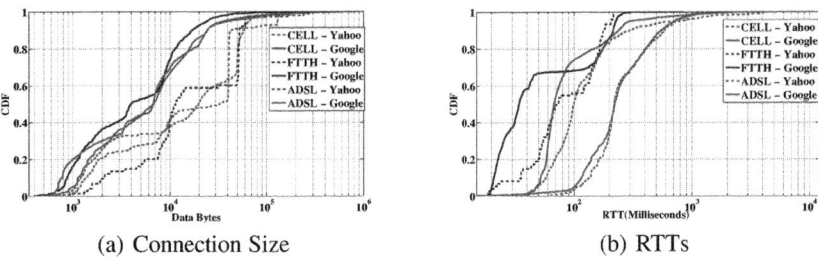

(a) Connection Size   (b) RTTs

Fig. 7. Yahoo vs. Google Web search services

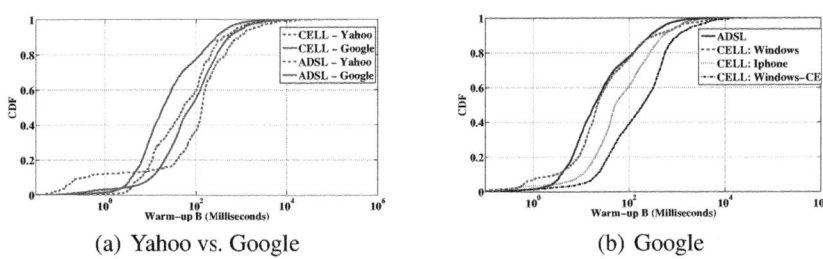

(a) Yahoo vs. Google   (b) Google

Fig. 8. Warm-up B

Figure 7(b) plots cdfs of RTTs. We can observe that RTT values on each access technology are similar for the two services, which suggests that the servers are located in France and that it is the latency of the first hop that dominates.

We do not present clustering results for Yahoo due to the small number of samples we have. However, a preliminary inspection of those results revealed the existence of clusters due to long Warm-up A values, i.e. long waiting times at the client side – in line with our observations with the Google Web search service. In the next section, we focus on the waiting time at the server side.

6.2 Data Preparation Time at the Server Side

Figure 8(a) presents the cdf of warm-up B[8] values for both Yahoo and Google for the ADSL and Cellular technology (we do not have enough samples on FTTH for Yahoo to present them). We observe an interesting result: for both Yahoo and Google, the time to craft the answer is longer for cellular than for the ADSL technology. It suggests that both services adapt the content to cellular clients. A simple way to detect that the remote client is behind a wired or wireless access is to check its Web browser-User Agent as reported in the HTTP header. This is apparently what Google does as Figure 8(b) reveals (again, due to a low number of samples on Yahoo, we are not able to report a similar breakdown). Indeed, cellular clients featuring a laptop/desktop Windows operating system (Vista/XP/2000) experience similar warm-up B as ADSL clients while clients using Iphones or a Windows-CE operating system experience way higher warm-up B. As the latter category (esp. Iphones: more than 66% of Google connections) dominates in our

---

[8] We have one total warm-up B value per connection, which is the total observed warm-up B for each train.

dataset they explain the overall cellular plot of Figure 8(a). Note that further investigations would be required to fully validate our hypothesis of content adaptation. We could think of alternative explanations like a different load on the servers at the capture time or some specific proxy in the network of the ISP. However, it is a merit of our approach to pinpoint those differences and attribute them to some specific components like the servers here.

7   Conclusion

In this paper, we tackled the issue of comparing networking applications over different access technology – FTTH, ADSL and Cellular. We focused on the specific case of Web search services. We showed that packet loss, latency, and the way clients interact with their mobile phones all have an impact on the performance metrics on the three technologies. We devised a technique that (i) automatically extracts the impact of each of these factors from passively observed TCP transfers and (ii) group together, with an appropriate clustering algorithm, the transfers that have experienced similar performance over the three access technology. Application of this technique to the Google Web search service demonstrated that it provides easily interpretable results. It enables for instance to pinpoint the impact of usage or of raw characteristics of the access technology. We further compared Yahoo and Google Web search traffic and provided evidences that they are likely to adapt content to the terminal capability for cellular clients which impacts the performance observed. As future work, we will apply our approach to the profiling of other network services, which should be straightforward since our approach is application agnostic (we did not make any hypothesis on Google Web search to profile it). We intend to profile, among others, applications which are more bandwidth demanding like HTTP streaming. We also would like to investigate the usefulness of the method at higher levels of granularity, such as session or client level.

Acknowledgments

This work has been partly supported by France Telecom, project CRE-46142809 and by European FP7 COST Action IC0703 Traffic Monitoring and Analysis (TMA).

References

1. Vu-Brugier, G.: Analysis of the impact of early fiber access deployment on residential internet traffic. In: International Teletraffic Congress, Paris (June 2009)
2. Pathak, A., Wang, Y., Huang, C., Greenberg, A., Hu, Y., Li, J., Ross, K.: Measuring and evaluating tcp splitting for cloud services
3. Svoboda, P., Ricciato, F., Keim, W., Rupp, M.: Measured web performance in gprs, edge, umts and hsdpa with and without caching. In: IEEE International Symposium on a World of Wireless, Mobile and Multimedia Networks, Helsinki, pp. 1–6 (June 2007)
4. Plissonneau, L., Vu-Brugier, G.: Mobile data traffic analysis: How do you prefer watching videos? In: ITC (2010)
5. Kammenhuber, N., Luxenburger, J., Feldmann, A., Weikum, G.: Web search clickstreams. In: Proceedings of the 6th ACM SIGCOMM Conference on Internet Measurement, IMC 2006, pp. 245–250. ACM, New York (2006)

6. Stamou, S., Kozanidis, L.: Impact of search results on user queries. In: Proceeding of the Eleventh International Workshop on Web Information and Data Management, WIDM 2009, pp. 7–10. ACM, New York (2009)

7. Padhye, J., Firoiu, V., Towsley, D.F., Kurose, J.F.: Modeling tcp throughput: A simple model and its empirical validation. SIGCOMM (1998)

8. Baccelli, F., McDonald, D.R.: A stochastic model for the throughput of non-persistent tcp flows. Perform. Eval. 65(6-7), 512–530 (2008)

9. Maier, G., Feldmann, A., Paxson, V., Allman, M.: On dominant characteristics of residential broadband internet traffic. In: Internet Measurement Conference, pp. 90–102 (2009)

10. Romirer-Maierhofer, P., Ricciato, F., D'Alconzo, A., Franzan, R., Karner, W.: Network-wide measurements of tcp rtt in 3g. In: Papadopouli, M., Owezarski, P., Pras, A. (eds.) TMA 2009. LNCS, vol. 5537, pp. 17–25. Springer, Heidelberg (2009)

11. Jaiswal, S., Iannaccone, G., Diot, C., Kurose, J., Towsley, D.: Measurement and classification of out-of-sequence packets in a tier-1 ip backbone. IEEE/ACM Trans. 15(1), 54–66 (2007)

12. Jiang, H., Dovrolis, C.: Passive estimation of tcp round-trip times. SIGCOMM Comput. Commun. Rev. 32(3), 75–88 (2002)

13. Shakkottai, S., Srikant, R., Brownlee, N., Broido, A., Claffy, K.C.: The rtt distribution of tcp flows in the internet and its impact on tcp-based flow control. Technical report number tr-2004-02, CAIDA (January 2004)

14. Veal, B., Li, K., Lowenthal, D.K.: New methods for passive estimation of TCP round-trip times. In: Dovrolis, C. (ed.) PAM 2005. LNCS, vol. 3431, pp. 121–134. Springer, Heidelberg (2005)

15. Zhang, Y., Breslau, L., Paxson, V., Shenker, S.: On the characteristics and origins of internet flow rates. In: SIGCOMM 2002, Pittsburgh, pp. 309–322 (2002)

16. Barbuzzi, A., Boggia, G., Grieco, L.A.: Desrto: an effective algorithm for srto detection in tcp connections. In: Ricciato, F., Mellia, M., Biersack, E. (eds.) TMA 2010. LNCS, vol. 6003, pp. 87–100. Springer, Heidelberg (2010)

17. Hafsaoui, A., Collange, D., Urvoy-Keller, G.: Revisiting the performance of short tcp transfers. In: 8th International IFIP-TC 6 Networking Conference, Aachen, pp. 260–273 (May 2009)

18. http://homepage.tudelft.nl/19j49/t-SNE.html

19. http://www.mathworks.com/help/toolbox/stats/kmeans.html

A Hadoop-Based Packet Trace Processing Tool*

Yeonhee Lee, Wonchul Kang, and Youngseok Lee

Chungnam National University
Daejeon, 305-764, Republic of Korea
{yhlee06,teshi85,lee}@cnu.ac.kr

Abstract. Internet traffic measurement and analysis has become a significantly challenging job because large packet trace files captured on fast links could not be easily handled on a single server with limited computing and memory resources. Hadoop is a popular open-source cloud computing platform that provides a software programming framework called MapReduce and the distributed filesystem, HDFS, which are useful for analyzing a large data set. Therefore, in this paper, we present a Hadoop-based packet processing tool that provides scalability for a large data set by harnessing MapReduce and HDFS. To tackle large packet trace files in Hadoop efficiently, we devised a new binary input format, called `PcapInputFormat`, hiding the complexity of processing binary-formatted packet data and parsing each packet record. We also designed efficient traffic analysis MapReduce job models consisting of map and reduce functions. To evaluate our tool, we compared its computation time with a well-known packet-processing tool, CoralReef, and showed that our approach is more affordable to process a large set of packet data.

1 Introduction

Internet traffic measurement and analysis needs a lot of data storage and high-performance computing power to manage a large traffic data set. Tcpdump [1] is widely used for capturing and analyzing packet traces, and various tools based on the packet capture library, "libpcap" [1], such as wireshark [2], CoralReef [3], and snort [4] have been deployed to handle packets. For aggregated information of a sequence of packets sharing the same fields like IP addresses and port numbers, Cisco NetFlow [5] is extensively employed to observe traffic passing through routers or switches in the unit of a flow. However, these legacy traffic tools are not suited for processing a large data set of tera- or petabytes monitored at high-speed links. In analyzing packet data for a large-scale network, we often have to handle hundreds of giga- or terabyte packet trace files. When the outbreaks of global Internet worms or DDoS attacks occur, we also have to process quickly

---

* This research was partly supported by the MKE (Ministry of Knowledge Economy), Korea, under the ITRC (Information Technology Research Center) support program supervised by the NIPA (National IT Industry Promotion Agency) (NIPA-2011-(C1090-1031-0005)) and partly by the IT R&D program of MKE/KEIT [KI001878, "CASFI : High-Precision Measurement and Analysis Research"].

J. Domingo-Pascual, Y. Shavitt, and S. Uhlig (Eds.): TMA 2011, LNCS 6613, pp. 51–63, 2011.
© Springer-Verlag Berlin Heidelberg 2011

a large volume of packet data at once. Yet, with legacy packet processing tools running on a single high-performance server, we cannot perform fast analysis of large packet data. Moreover, with a single-server approach, it is difficult to provide fault-tolerant traffic analysis services against a node failure that often happens when intensive read/write jobs are frequently performed on hard disks.

MapReduce [6], developed by Google, is a software paradigm for processing a large data set in a distributed parallel way. Since Google's MapReduce and Google file system (GFS) [7] are proprietary, an open-source MapReduce software project, Hadoop [8], was launched to provide similar capabilities of the Google's MapReduce platform by using thousands of cluster nodes. Hadoop distributed filesystem (HDFS) is also an important component of Hadoop, that corresponds to GFS. Yahoo!, Amazon, Facebook, IBM, Rackspace, Last.fm, Netflix and Twitter are using Hadoop to run large-scale data-parallel applications coping with a large set of data files. Amazon provides Hadoop-based cloud computing services called Elastic Compute Cloud (EC2) and Simple Storage Service (S3). Facebook also uses Hadoop to analyze the web log data for its social network service. From the cloud computing environment like Hadoop, we could benefit two features of distributed parallel computing and fault tolerance, which could fit well for packet processing tools dealing with a large set of traffic files. With the MapReduce programming model on inexpensive commodity PCs, we could easily handle tera- or petabyte data files. Due to the cluster filesystem, we could provide fault-tolerant services against node failures.

In this paper, hence, we present a Hadoop-based packet trace processing tool that stores and analyzes the packet data on the cloud computing platform. Major features of our tool are as follows. First, it could write packet trace files in libpcap format on HDFS, with which the problem of archiving and managing large packet trace files is easily solved. Second, our tool could significantly reduce the traffic statistics computation time of large packet trace files with MapReduce-based analysis programs, when compared with the traditional tool. For these purposes, we have implemented a new binary file input/output module, PcapInputFormat, which reduces the processing time of packet records included in trace files, because text files are used for the conventional input file format in Hadoop. In addition, we have designed packet processing MapReduce job models consisting of map and reduce tasks.

The remaining of this paper is organized as follows. In Section 2, we describe the related work on MapReduce and Hadoop. The architecture of our tool and its components are explained in Section 3, and the experimental results are presented in Section 4. Finally Section 5 concludes this paper.

2   Related Work

For Internet traffic measurement and analysis, there are a lot of packet-processing tools such as tcpdump, CoralReef, wireshark, and snort. Most of these packet-processing tools are run on a single host with the limited computing and storage resources.

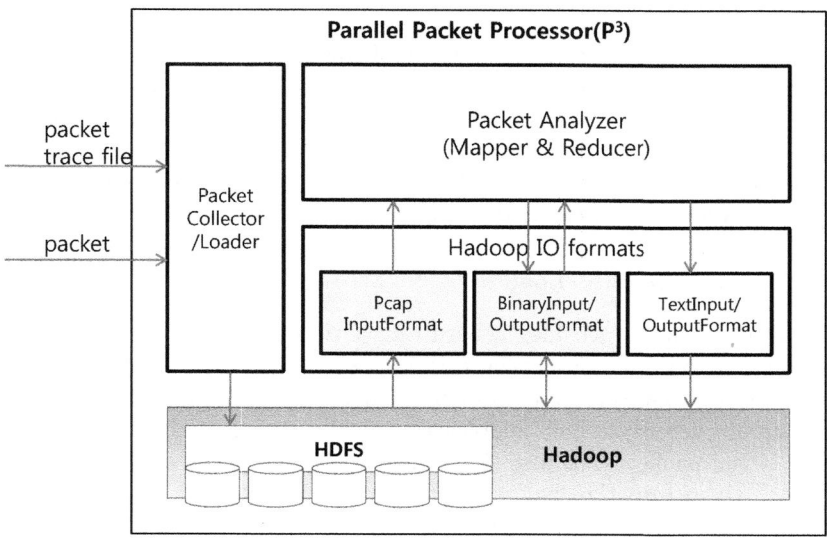

Fig. 1. The overview of Hadoop-based packet file processing tool, Parallel Packet Processor (P^3)

As the MapReduce platform has been popular, a variety of data mining or analytics applications are emerging in the fields of natural sciences or business intelligence. Typical studies based on Hadoop are text-data analysis jobs like web indexing or or log analysis. For the network management fields, snort log analysis was tried with Hadoop in [9]. In our previous work [10], we have devised a simple MapReduce-based NetFlow analysis method that could analyze text-converted NetFlow v5 files, which showed that MapReduce-based NetFlow analysis outperforms popular flow-tools. In this paper, we present a Hadoop-based packet trace analysis tool that could process enormous volumes of libpcap packet trace files.

On the other hand, there has been few work on dealing with non-text files coherently in Hadoop. As for extending the Hadoop API, Conner [11] has customized Hadoop's `FileInputFormat` for image processing, but has not clearly described its performance evaluation results. Recently, there have been a few studies [12, 13, 14] to improve the performance of Hadoop. Zaharia *et al.* [12] have designed an improved scheduling algorithm that reduces Hadoop's response time by considering a cluster node performing poorly. Kambatla *et al.* [13] proposed a Hadoop provisioning method by analyzing and comparing resource usage patterns of various MapReduce jobs. Condie *et al.* [14] devised a modified MapReduce architecture that allows data to be pipelined between nodes, which is useful for interactive jobs or continuous query processing programs.

3   Processing Packet Trace Files with Hadoop

Parallel Packet Processor (P^3)[1] consists of three main components: the *Packet Collector/Loader* that saves packets to HDFS from the online packet source using libpcap or the captured trace file; the *Packet Analyzer* implemented by map and reduce tasks processing packet data; and the *Hadoop IO formats* that read packet records from files on HDFS and return the analysis results. The overall architecture is illustrated in Fig. 1.

3.1   Packet Collector/Loader

P^3 analyzes packet trace files generated by either a packet sniffing tool such as tcpdump or our own packet-capturing module, *Packet Collector/Loader*. The *Packet Collector* captures packets from a live packet stream and save them to HDFS simultaneously, which enables P^3 to collect packets directly from packet probes. Given packet trace files in the libpcap format, *Packet Loader* reads files and splits them into the fixed-size of chunks, and save the chunks of files to HDFS. Network operators will irregularly copy captured packet trace files to HDFS in order to analyze detailed characteristics of the trace.

The *Packet Collector* uses libpcap and jpcap modules for capturing packets from a live source, and the HDFS stream writer for saving packet data. To use this stream writer, we developed a HDFS writing module in Java. The *Packet Collector* can capture packets by interacting with the libpcap module through jpcap. The *Packet Loader* just saves the packet trace file to HDFS by using the Hadoop stream writer. Within HDFS, each packet trace file larger than the specified HDFS block size (typically 64 MB) is split, and three replicas (by default) for every block are copied into HDFS for fault tolerance. The trace file contains binary-formed packet data, which has the non-fixed length of packet records.

3.2   New Binary Input/Output Format for Packet Records

In the Hadoop cluster environment, network bandwidth between nodes is a critical factor for the performance. To overcome the network bottleneck, MapReduce collocates the data with its computing node if possible. HDFS has concepts of a block which is the unit for writing a file and a split which is the unit for being processed by MapReduce task. As a block is in a large unit (64 MB by default), a file in HDFS will be divided into block-sized chunks that are stored as independent units. A split is a chunk of the input file that is processed by a single map task. Each split is divided into records, and passed to the map to process each record of a key-value pair in turn by an `InputFormat`.

The text file is a common format for input/output in Hadoop. `TextInputFormat` can create splits for the text file and parse each split into each line of records by carriage return. However, packet trace files are stored in the binary format defined by libpcap and have no carriage return. Thus, it is necessary to convert binary packet

---

[1] Source codes are available in [15].

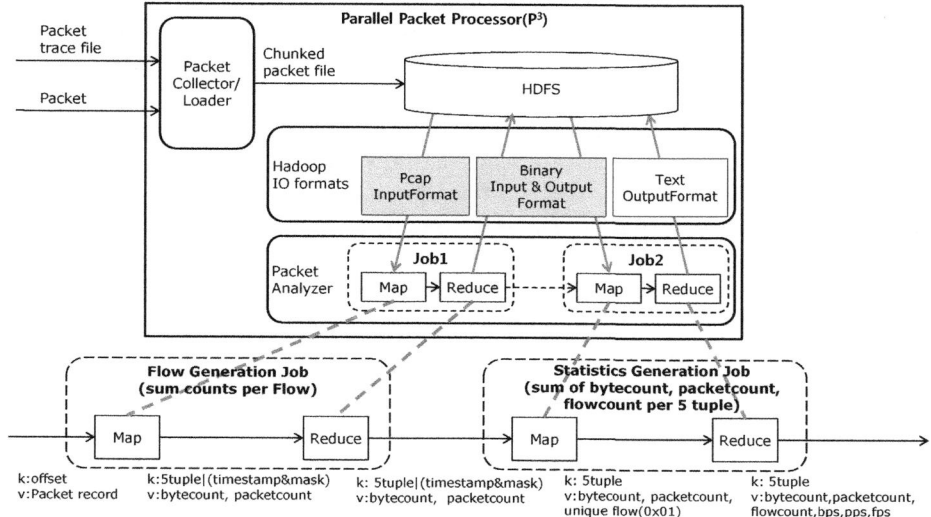

Fig. 2. New Hadoop input/output formats for processing packet trace. This example shows how the periodic flow statistics is computed. `PcapInputFormat` is used for reading packet records to generate flows from packet trace files. `BinaryOutputFormat` is for writing binary-form of flow records to HDFS, and `BinaryInputFormat` for reading flow records from HDFS to generate flow statistics.

trace files to text ones for the input to Hadoop using `TextInputFormat`. This conversion requires sequential processing which consists of reading variable-length of binary packet records by using the packet length field specified in each packet header, converting them to text-formed ones, and writing to text ones. Thus, this causes additional computing time and storage resources, and does not correspond with a parallel manner of Hadoop.

The process of a Hadoop job consists of several steps. First, before starting a job, we save large packet trace files on HDFS cluster nodes in the unit of fixed-length of blocks. When a job is started, the central job controller, `Jobtracker`, assigns map tasks to each node to process blocks, and then, map tasks in each node reads records from the assigned block in parallel. At this point, both the start location and end location of `InputSplits` can not be clearly detected within a fixed-size HDFS block, because each packet record has variable length of binary data without any record boundary character like a carriage return. That is, there is no way for each map to know the location of the staring position of the first packet in the block until another map task running on the previous block finishes reading the last packet record.

In order to solve this boundary detection issue in manipulating the binary packet trace files on HDFS in parallel, we have devised `PcapInputFormat` as a new packet input module. `PcapInputFormat` could support parallel packet processing by finding the first and last packet location in the HDFS blocks, and by parsing variable-length of packet records. We also devised `BinaryInputForamt` as the input module of fixed-sized binary records, and `BinaryoutputFormat`, the output module of fixed-sized binary records to handle flow statistics using packet data. Figure 2 shows how the native libpcap binary files are processed with binary packet input/output formats in the modified Hadoop. There are two continuous jobs. In the first job, `PcapInputFormat` is used for reading packets from packet trace files to generate flows. `BinaryOutputFormat` is responsible for writing binary-form of flow records to HDFS. In the second job, we compute the flow statistics with `BinaryInputFormat` by reading flow records. Hadoop in itself could support binary data as inputs and outputs with sequence file format, but we have to transform packet traces to sequence files by parsing every packet record sequentially. Therefore, it will not be efficient to use the built-in sequence file format for handling packet trace files.

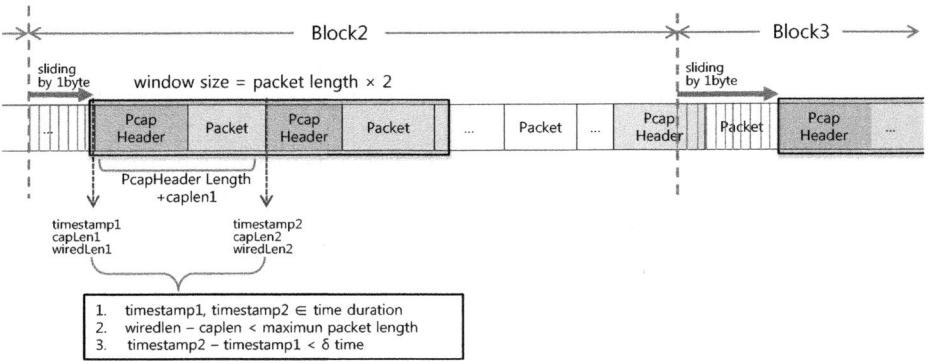

Fig. 3. Our heuristic method of `PcapInputForamt` for finding the first record from each block using sliding-window

PcapInputFormat. `PcapInputFormat` includes a sub-class, called `PcapVlenRecordReader`, that reads packet records from a block on Hadoop, and it is responsible for parsing records from the split across blocks. When each map task reads packet records from blocks stored on HDFS through the `InputFormat` concurrently, `PcapInputFormat` has to find the first packet in the assigned block, because the last packet record in the previous block might be stored across two continuous HDFS blocks. However, it is not easy to pinpoint the position of the first packet in a block without searching records sequentially due to variable packet size without boundary character of each packet record. Hence, how to identify the boundary of a split from distributed blocks is a main concern of MapReduce tasks for packet processing. That is, the map task has to know the starting position and ending position of splits with only partial fragments of the trace files. Thus, we employ a heuristic to find the first packet of each block by searching a timestamp field included in the packet, assuming that the timestamps of two continuous packets stored in the libpcap trace file will not be much different. The threshold of timestamp difference might be configured to consider the characteristics of packet traces on various link speeds.

The libpcap packet header consists of four kinds of byte arrays: *timestamp of seconds, timestamp of microseconds, captured length,* and *wired length. captured length* is the number of octets of a packet saved in the file and *wired length* is the actual length of a packet record flowed into. Our heuristic to find the first/last packet records needs two input timestamp parameters, which limits a permissible range of *timestamp* for each packet.

Using these values, we perform a sliding-window pattern matching algorithm with the candidate timestamp field on the block. Figure 3 illustrates this method. First, `PcapVlenRecordReader` takes in a sample byte window of 2 × the maximum packet length from the beginning portion of the block. To find the beginning point of the split, we suppose that first four bytes consist of the *timestamp* of the first packet. According to pcap header format, another four bytes later, it might be followed by four bytes of *captured length* and *wired length* fields. This assumption let us know that the next packet is followed by the length of bytes as noticed in *captured length* of the first packet. Thus, we can extract its header information for validation.

To validate the assumption that the first four-byte field represents a timestamp of the first packet in an assigned block, we conduct three intuitive verification procedures. First, we make sure that the first timestamp and the second one are within the time duration for capturing packets given by user. Second, we examine that the gap between *captured length* and *wired length* of each packet is smaller than the maximum packet length. Third, we investigate that the gap between the first timestamp and the second one is less than δ *time*, which might be acceptable as an elapsed time for two consecutive packets. We set this value as 60 by default. `PcapVlenRecordReader` repeats this pattern matching function by moving a single byte until it finds the position that meets these conditions.

An extracted record is a byte array that contains the binary-formed packet information and it is passed to the map task as a key-value pair of the `LongWritable` byte offset of the file and a `BytesWritable` byte array of a packet record. Through experiments of a large data set, we confirmed that this practical heuristic works well and the performance is affordable at the same time.

BinaryInputFormat and BinaryOutputFormat. The `PcapInputFormat` module in the modified Hadoop manages libpcap files in a binary format. The *Packet Analyzer* generates basic statistics such as the total IPv4 byte count or packet count. Besides, it also computes periodic statistics results such as bit rate, packet rate, or flow rate for a fixed-length of a time interval. To create these statistics, we have to produce flows from raw packets. In Fig. 2, `PcapInputFormat` is used for the first MapReduce job as a carrier of variable length of packet records from packet trace files. After flows are formulated at the first MapReduce job, they are stored to HDFS through the `BinaryOutputFormat` in the fixed-length of byte arrays. To calculate the statistics of these flows, we deliver flows to the next MapReduce job with `BinaryInputFormat`.

`BinaryInputFormat` and `BinaryOutputFormat` need one parameter, the size of the record to read. In the case of the periodic flow statistics computation job, the input parameter will be the size of a flow record. `BinaryInputFormat` parses a flow record of a byte array, and passes it to the map task as a key-value pair of a `LongWritable` byte offset of the file and a `BytesWritable` byte array of a flow record. Likewise, `BinaryOutputFormat` saves the record consisting of a `BytesWritable` key and a `BytesWritable` value produced by the job to HDFS. Our new binary input/output formats could enhance the speed of loading the libpcap packet trace files on HDFS, and reading/writing packet records in the binary file format in HDFS.

3.3   Packet Analyzer

For the *Packet Analyzer*, we have developed four MapReduce analysis commands.

Total traffic statistics. First, we compute the total traffic statistics such as byte/packet/flow counts for IPv4/v6/nonIP. As shown in Fig. 4, the first MapReduce job calculates the total traffic information. In order to count the number of unique IP addresses and ports, we need another MapReduce job. For this purpose, the first job emits a new key-value pair consisting of a key combined with the text name and an IP address, and a value of 1. This key will be used for identifying unique IP address or port records. The second job summarizes the number of unique IP addresses and ports.

Periodic flow statistics. Periodically, we often assess each flow information consisting of 5-tuples of IP addresses and ports from packet trace files. For flow analysis, we have implemented two MapReduce jobs for periodic flow statistics and aggregated flow information, respectively. As shown in Fig. 5, the first job computes the basic statistics for each flow during the time interval. Then, the second job will aggregate the same flows lasting longer than the small time interval into a single flow. Thus, the first job emits a new key-value pair for the aggregated flows. The key consists of the 5-tuple text concatenated by the masked timestamp.

Periodic simple statistics. A simple statistics is to tally the total byte/packet and bit/packet rate per each interval, which could be implemented with a single MapReduce job. During the map phase, a simple statistics job classifies packets as non-ip, IPv4, and IPv6, and creates a new key with timestamp. The timestamp for the new key is masked by the interval value to produce periodic statistics.

Top N statistics. Given traffic traces, we are interested in finding the most popular statistics such as top 10 flow information. The `Top N` module makes full use of MapReduce features to solve this purpose. We create a new key of the record for identifying or grouping records within the map phase. Thus, in the reduce phase, all the records hashed by keys become sorted by the build-in MapReduce sorting function. Therefore, the reduce task just emits specified number of record data. Figure 6 shows the process for computing top N information. The input could be the results of periodic simple statistics job or periodic

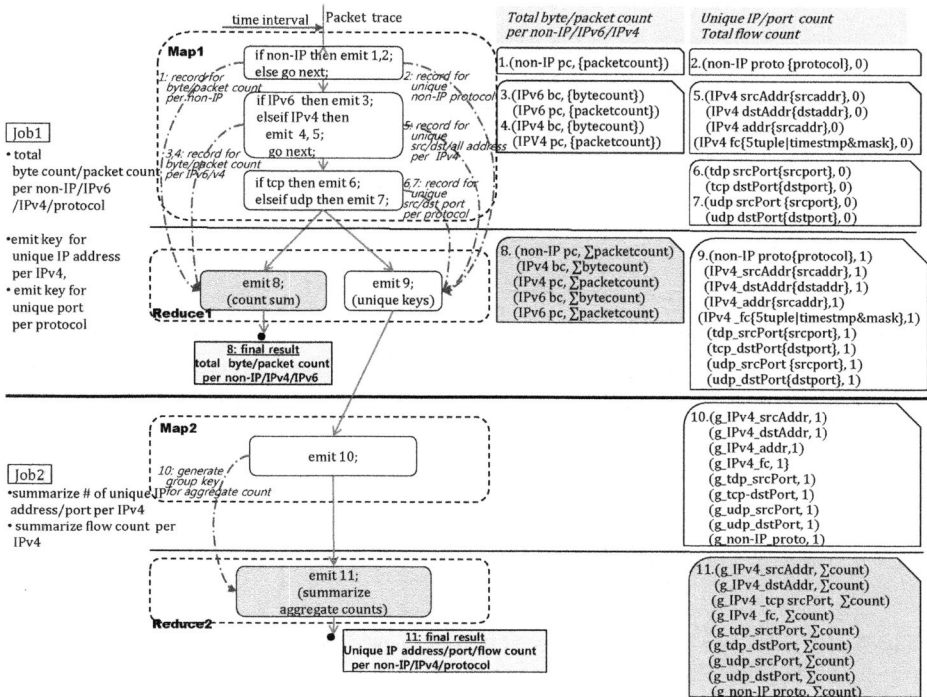

Fig. 4. Total traffic statistics (total byte/packet/flow count per IPv4/v6/non-IP and the number of unique IP addresses/ports): two MapReduce jobs are necessary to count the number of hosts and ports

flow statistics job. Thus, the `Top N` module will add one more MapReduce job. The `Top N` module requires two parameters: one is the column name to sort and the other is the number, N, to output. The column name can be byte count, packet count, and flow count. At the map task, the column is used for creating a new key which will be used for sorting in the running shuffle and sort phase by reduce task. The reduce task just emits records from top to N^{th}.

4   Experiments

For experiments, we have setup standard and high-performance Hadoop testbeds in our laboratory (Table 1). A standard Hadoop testbed consists of a master node and four slave nodes. Each node has quad-core 2.83 GHz CPU, 4 GB memory, and 1.5 TB hard disk. All Hadoop nodes are connected with 1 Gbps Ethernet cards. For the comprehensive test with large files, we configured a high-performance testbed of 10 slave nodes that have octo-core 2.93 GHz Intel i7 CPU, 16 GB memory, 1 TB hard disk, and 1 Gbps Ethernet card. We used Hadoop 0.20.3 version with the HDFS block size of 64 MB. The replication factor of standard/high-performance testbeds is two/three. For comparing our

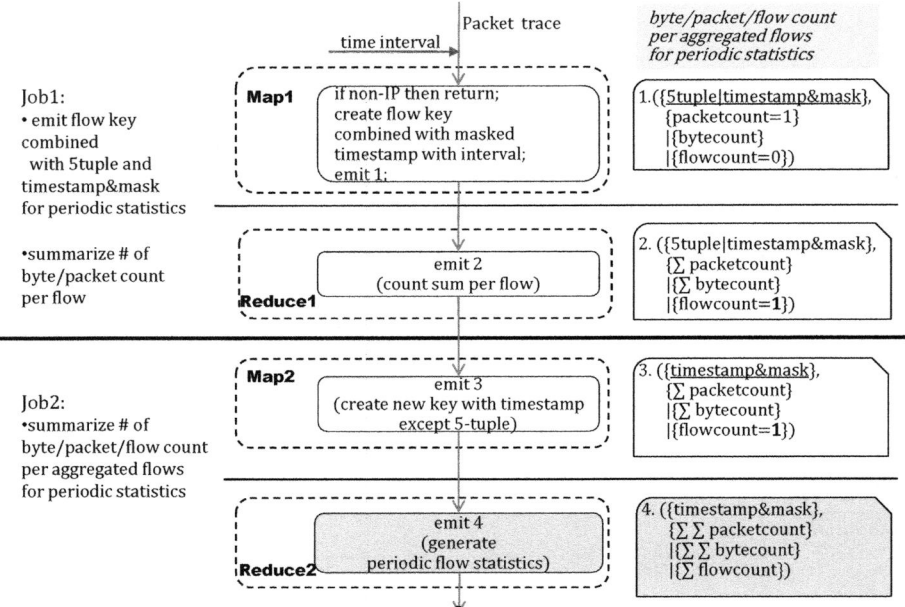

Fig. 5. Periodic flow statistics (flow statistics regarding byte/packet/flow counts per time window): two MapReduce jobs are necessary to perform the aggregated flow statistics

tool with CoralReef, we also have configured a single node that has the same hardware specification with a node of the standard Hadoop testbed. Table 2 shows datasets we used for our experiments. We performed experiments with various amount of datasets from 10, 100, 200, and 400 GB.

Table 1. P^3 testbed

Type	Nodes	CPU	Memory	Hard Disk
Single CoralReef node	1	2.83 GHz (Quad-core)	4 GB	1.5 TB
Standard Hadoop Testbed	5	2.83 GHz (Quad-core)	4 GB	1.5 TB
High-performance Hadoop Testbed	10	2.93 GHz (Octo-core)	16 GB	1 TB

4.1   Scalability

For the scalability test, we ran total traffic statistics and periodic simple statistics jobs under various file sizes. In order to know the impacts of Hadoop resources on performance, we executed P^3 on two Hadoop testbeds of standard 4 nodes and high-performance 10 nodes. For the comparison, we measured the computation time of CoralReef on the standard node. Figure 7 depicts the average computing time for the total traffic statistics (total byte/packet/flow count per IPv4/IPv6/non-IP and the number of unique IP addresses/ports) when the file size varies from 10 to 400 GB. We observed that job completion time is directly proportional to the data volume to be processed by tools. Overall, P^3 on 10

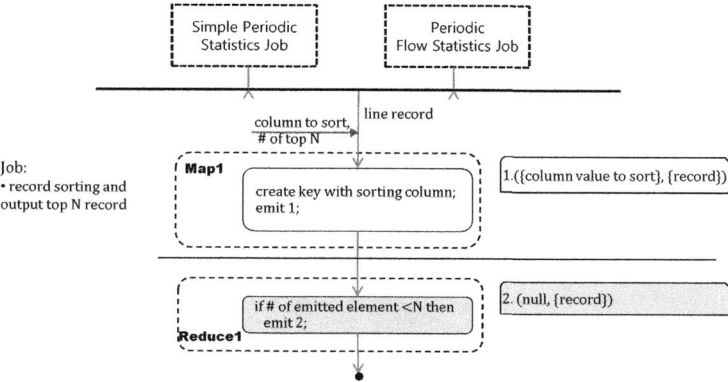

Fig. 6. Top N statistics: the top N number of records of periodic flow statistics or periodic simple statistics

Table 2. Test packet trace files

Type	# of packet files	# of packets
10 GB	1	9.4 M
100 GB	1	92.7 M
200 GB	2	185.4 M
400 GB	7	441.1 M

high-performance Hadoop nodes ($P^3(H, 10)$) shows the better performance than CoralReef as well as P^3 on four standard Hadoop workers, $P^3(S, 4)$. It is shown that $P^3(S, 4)$ does not much improve the performance than CoralReef, because the packet processing MapReduce job spends its time in reading/writing data with disk I/O's more than in performing the computation task. However, in case of high-performance 10 worker nodes, P^3 outperforms CoralReef. In 200/400 GB data, $P^3(H, 10)$ completes the job 6.2 (5.8) × faster than CoralReef.

Next, we conducted an experiment with the periodic simple statistics command. In Fig. 8, it is seen that P^3 on 10 Hadoop nodes ($P^3(H, 10)$) finishes the computation job faster than CoralReef and $P^3(S, 4)$. In 200/400 GB, compared with CoralReef, $P^3(H, 10)$ achieved 7.5 (7.3) × performance improvement. Though the speed-up ratio is the maximum at 200 GB, we could still reduce the computation time at 400 GB with $P^3(H, 10)$. From the experiments, it is seen that resource-proportional computing could be possible for a large data input with P^3.

4.2   Observation

From the experiments, we could find an interesting observation that Hadoop is especially useful for computing-intensive jobs. We compared the total traffic statistics and periodic flow statistics commands. The total traffic statistics command is more complex than the periodic flow statistics command, because it calculates 14 different analysis items. Therefore, we first expected that the

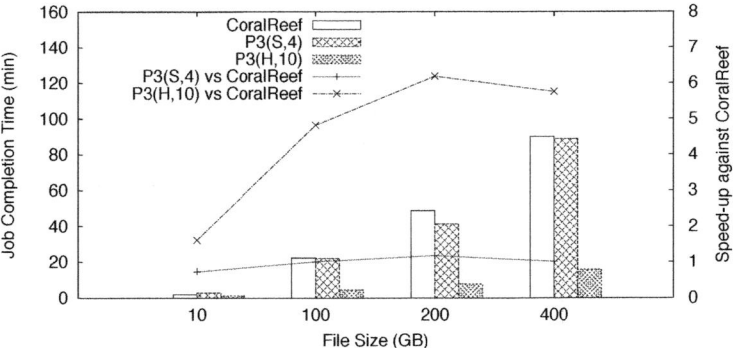

Fig. 7. Completion time of a total traffic statistics job (total byte/packet/flow count per IPv4/v6/non-IP and the number of unique IP addresses/ports): CoralReef vs. P^3 regarding various file sizes and the speed-up

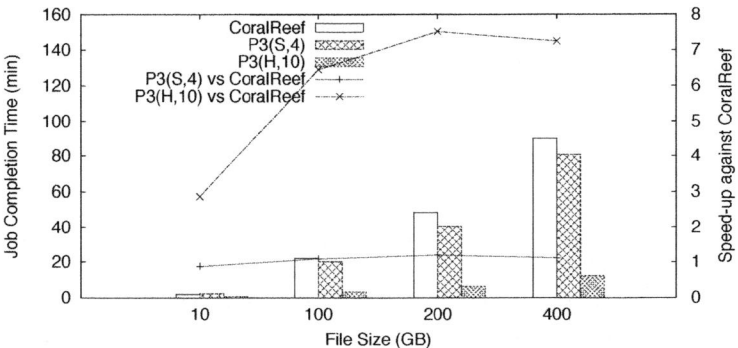

Fig. 8. Completion time of a periodic simple statistics job (byte/packet count per time window): CoralReef vs. P^3 regarding various file sizes and the speed-up

Table 3. Comparison of total traffic statistics and periodic flow statistics tools under 400 GB

	Total Traffic Statistics	Periodic Flow Statistics
# of MapReduce jobs	2	2
# of statistics items	14	2
# of intermediate records by map	3,961 M	441 M
# of intermediate bytes by map	122 GB	16 GB
Completion time (sec)	939	798

performance of the flow statistics command is much better than the total traffic statistic command. However, given the same input data of 400 GB, the completion time of the total traffic statistics command is longer than that of the flow statistics command only by 18%, even though that job emits 9 × more intermediate records from the map phase as shown in Table 3. This result implies that packet processing and analyzing jobs are I/O-intensive and I/O processing from/to HDFS is the bottleneck of the performance. Therefore, we have to integrate multiple computation jobs together while reducing I/O times in order to develop high-performance traffic analysis applications in MapReduce.

5   Conclusion

We have presented a scalable Hadoop-based parallel packet processor that could analyze large packet trace files. Compared to the conventional packet tools such as tcpdump or CoralReef, our proposal could easily manage large packet trace files of tera- or petabytes, because we have employed the MapReduce platform for parallel processing. In addition, due to the distributed filesystem, HDFS, we could provide the fault tolerance to our traffic analysis system. For packet analysis, we designed representative statistics computation modules with MapReduce. From the experiments, we have shown that our tool outperforms a typical traffic analysis method running on a single server. For the future work, we plan to optimize the performance of MapReduce jobs, to extend P^3 for real-time packet analysis under high-speed links and to enhance the pattern matching algorithm for parsing packet records.

References

1. Tcpdump, http://www.tcpdump.org
2. Wireshark, http://www.wireshark.org
3. CAIDA CoralReef Software Suite,
   http://www.caida.org/tools/measurement/coralreef
4. Roesch, M.: Snort - Lightweight Intrusion Detection for Networks. In: USENIX LISA (1999)
5. Cisco NetFlow, http://www.cisco.com/web/go/netflow
6. Dean, J., Ghemawat, S.: MapReduce: Simplified Data Processing on Large Cluster. In: OSDI (2004)
7. Ghemawat, S., Gobioff, H., Leung, S.: The Google file system. In: SOSP (October 2003)
8. Hadoop, http://hadoop.apache.org/
9. Chen, W., Wang, J.: Building a Cloud Computing Analysis System for Intrusion Detection System. In: CloudSlam (2009)
10. Lee, Y., Kang, W., Son, H.: An Internet Flow Analysis Method with MapReduce. In: 1st IFIP/IEEE Workshop on Cloud Management (April 2010)
11. Conner, J.: Customizing Input File Formats for Image Processing in Hadoop, Arizona State University Technical Report (2009)
12. Zaharia, M., Konwinski, A., Joseph, A.D., Katz, R., Stoica, I.: Improving MapReduce Performance in Heterogeneous Environments. In: OSDI (2008)
13. Kambatla, K., Pathak, A., Pucha, H.: Towards Optimizing Hadoop Provisioning in the Cloud. In: USENIX Hotcloud (2009)
14. Condie, T., Conway, N., Alvaro, P., Hellerstein, J.M., Elmeleegy, K., Sears, R.: MapReduce Online. ACM/USENIX NSDI (April 2010)
15. https://sites.google.com/a/networks.cnu.ac.kr/yhlee/p3

Monitoring of Tunneled IPv6 Traffic Using Packet Decapsulation and IPFIX (Short Paper)

Martin Elich[1,3], Matěj Grégr[1,2], and Pavel Čeleda[1,3]

[1] CESNET, z.s.p.o., Prague, Czech Republic
[2] Brno University of Technology, Brno, Czech Republic
`igregr@fit.vutbr.cz`
[3] Masaryk University, Brno, Czech Republic
{`elich,celeda`}`@mail.muni.cz`

Abstract. IPv6 is being deployed but many Internet Service Providers have not implemented its support yet. Most of the end users have IPv6 ready computers but their network doesn't support native IPv6 connection so they are forced to use transition mechanisms to transport IPv6 packets through IPv4 network. We do not know, what kind of traffic is inside of these tunnels, which services are used and if the traffic does not bypass security policy. This paper proposes an approach, how to monitor IPv6 tunnels even on high-speed networks. The proposed approach allows to monitor traffic on 10 Gbps links, because it supports hardware-accelerated packet distribution on multi-core processors. A system based on the proposed approach is deployed at the CESNET2 network, which is the largest academic network in the Czech Republic. This paper also presents several statistics about tunneled traffic on the CESNET2 backbone links.

Keywords: IPv6, Teredo, ISATAP, 6to4, network monitoring, IPv6 tunnel, IPFIX, FlowMon.

1 Introduction

End users have nowadays IPv6 ready computers, because support for this protocol is available in main operating systems (Windows, Linux, BSD, Mac OS X). Unfortunately, not every ISP has implemented IPv6 support yet, which together with IPv6 backward incompatibility with IPv4 protocol requires transition mechanisms. 6to4, Teredo and ISATAP are the most used transition techniques. These three methods use encapsulation of IPv6 protocol inside IPv4 protocol – tunneling. The encapsulation hides the IPv6 traffic. Tunneled traffic may look like ordinary IPv4 traffic using UDP ports, so administrators do not know, which IPv6 network service is requested, how much traffic flows through tunnels etc. IPv6 tunnels are created automatically so there is no need for a user intervention. This can cause security problems such as bypassing firewalls, unauthorized use of services etc.

We propose an approach how to overcome this limitation and how to monitor tunneled IPv6 traffic. It features hardware-accelerated packet distribution with

J. Domingo-Pascual, Y. Shavitt, and S. Uhlig (Eds.): TMA 2011, LNCS 6613, pp. 64–71, 2011.
© Springer-Verlag Berlin Heidelberg 2011

which it is possible to monitor even 10 Gbps links. Statistics and tunneled traffic distribution presented in this paper are generated from IPFIX data collected on CESNET2 backbone links, which is the largest academic network in the Czech Republic.

The paper is organized as follows. Section 2 describes related work. IPv6 transition techniques are described in Section 3. Proposal of architecture for monitoring tunneled data is in Section 4. Section 5 shows several statistics and analysis from network monitoring and Conclusion is in Section 6.

2   State-of-the-Art and Contribution

Several papers discuss and present IPv6 address and traffic analysis. Authors in [4] analyze traffic from a US Tier-1 ISP. Analyzed traffic in their data-set consists mainly of DNS and ICMP packets. They believe that it is because ISP's customers consider IPv6 traffic still as experimental. For IPv6 address assignment they used methodology introduced in [5]. Statistics from a China Tier-1 ISP are presented in [3]. Their observation about address assignment and application usage are similar to ours with some exceptions. Their traffic contains higher proportion of native IPv6 traffic. We believe, that it is due to larger expansion of IPv6 in China and Asia.

Unfortunately analysis of tunneled IPv6 traffic is missing in many papers. Some statistics are presented in [4] but just for Teredo traffic. Paper [6] observes IPv6 traffic on 6to4 relay but it is quite old. Despite our best efforts we did not find publications about tunneled IPv6 traffic in ISATAP tunnels. Statistics about 6to4 tunnels or Teredo are not so detailed and up to date. This paper tries to update knowledge about nowadays native and tunneled IPv6 traffic.

Contribution of this paper consists of several parts. First, we propose an approach, how to extend IPFIX to provide possibility to monitor tunneled IPv6 traffic. This approach is scalable and can be used in very large networks for monitoring IPv4, native IPv6 and tunneled IPv6 traffic. It is possible to use our concept to collect traffic on high-speed 10 Gbps links with no need to use packet sampling. Second, we present several statistics for tunneling mechanisms. Deployment of IPv6 protocol accelerates because new operating systems use this protocol by default. Therefore more services are accessible through IPv6 protocol and traffic distribution is nowadays completely different than before. Hence current statistics are very useful.

3   Transition Techniques

IPv6 connectivity is enabled and preferred in most operating systems by default. If a station is connected to local IPv4 network without native IPv6 connectivity and web site or another network service is accessible through both protocols, IPv6 has precedence and a host tries to communicate through this protocol first. Because IPv6 is not compatible with the previous IPv4 protocol, different types of transition techniques were proposed. The most interesting are tunneling techniques, because we do not know, which protocols and services are used inside the tunnels. 6to4, Teredo and ISATAP are todays most used tunneling mechanisms for connection to IPv6 network.

6to4 tunneling is the most used transition technique today. According to priority in operating system, if a network device has public IPv4 address, 6to4 is the first mechanism to be used. A host construct 64 bits long IPv6 network prefix according to rules described in [8]. Last 64 bits are used as EUI (End Unit Identifier). Several techniques can be used to create the identifier: based on EUI-64, manual assignment or randomly generated [2]. Default configuration in Windows or Linux use well-known EUI values in practice. Linux use the value 1 by default and Windows XP, Vista, 7 use IPv4 address in lower 32 bits of the EUI [12]. When sending packets, the 6to4 tunnel wraps an IPv6 datagram into an IPv4 datagram with protocol number 41.

Teredo was designed to be able to send network traffic through NAT [7]. It does not encapsulate IPv6 packet in protocol 41 but send it via UDP packet on default port 3544. Teredo address is more complicated then 6to4 and consists of Teredo prefix, Teredo server address, flags, port and client's external address. When simple encapsulation is used only the IPv6 packet is carried as the payload of an UDP packet. Server may insert other fields such as Origin and Authentication.

ISATAP – Intra-Site Automatic Tunnel Addressing Protocol is an IPv6 transition mechanism used in local networks to connect islands of IPv6 nodes over IPv4 networks. Connection to the Internet is made by another mechanism such as 6to4. ISATAP like 6to4 uses encapsulation in protocol number 41 [9]. Nowadays ISATAP is usually the last used transition techniques. Transition techniques order which a host tries when does not have native IPv6 connectivity is usually 6to4, Teredo, ISATAP.

4   Architecture and Implementation

The proposed approach for tunneled IPv6 traffic monitoring describes whole process of IP flow generation, export and collection. The flows are generated by FlowMon exporter a software probe which is able to export NetFlow and IPFIX data. The FlowMon exporter is able to generate flow statistics from any source if the input plug-in supports it [1].

4.1   Architecture

The proposed approach consist of three layers (see Figure 1). The first layer can be a network card or a more specialized hardware. Purpose of this layer is capturing packets and sending them over the software interface to the input plug-in. We used the FPGA based COMBOv2 card and libsze2 library as a software interface. We developed FPGA design for COMBOv2 cards HANIC (Hardware-Accelerated Network Interface Card) which provides a high precision timestamp

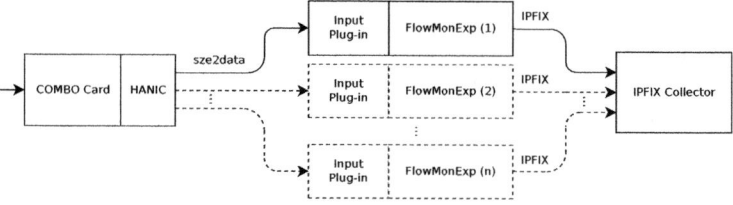

Fig. 1. System architecture – packets are captured by the COMBOv2 card and can be distributed to 16 FlowMon exporters with loaded input plug-in. IP flows are generated based on processed packets and later exported in IPFIX format.

generated for each packet. Packets can be distributed to several DMA (Direct Memory Access) channels. Packet distribution is one of benefits of proposed approach and is described in Section 4.2.

The second layer reads packets from the software interface and processes them with the FlowMon exporter [1]. We designed and implemented input plug-in for monitoring of IPv6 tunneled traffic but plug-ins can have any other functionality.

The plug-in for tunneled IPv6 traffic monitoring detects packets, which are part of tunnels, using a defined set of rules. After tunnel is detected, IPv4 header is stripped out and packets are processed by IPv6 header parser. Relevant information from packet are stored to a data structure representing a part of flow (in this case flow containing single packet). This filled data structure is passed to the exporter. More about plug-in functionality can be found in Section 4.3. The exporter generates flow statistics based on data structures from the input plug-in. Flow statistics are exported in IPFIX using custom IPFIX templates with enterprise-specific information elements to carry information about the tunnel.

The third and last layer is the IPFIX collector.

4.2 Packet Distribution

Packet distribution is implemented using the HANIC design. The goal is to distribute packets between several instances of the FlowMon exporter on the hardware level.

The HANIC design provides a packet header parser. The parser can extract necessary fields for flow identification. The output of parsing unit is a sequence of bits with fixed length of 301 bits. This sequence is then passed to the HASH unit which computes CRC hash with length of $log_2(number\ of\ channels)$. Each packet is send to one of channels according to its hash (the hash is used to address a channel).

Current version of the design use hash length of four bits. This allows to distribute packets to 16 instances of the FlowMon exporter without breaking the flow cache. Another advantage is possibility to process packets on multiple processors which greatly improves overall performance.

4.3   Plug-In Implementation

The input plug-in is implemented as shared library for Linux. It filters and preprocess each packet to data structure compatible with the FlowMon exporter plug-in API. The input plug-in reads packets from the COMBOv2 card in a form of memory chunks. These memory chunks consist of whole packet together with high precision timestamp and card's interface identifier from which packet was read. Protocol number is extracted from Ethernet header or from the MPLS label if MPLS is used.

All IPv4 packets are processed by the filters to detect presence of tunneling. Detection supports the following tunneling mechanisms: Teredo, 6to4 and ISATAP. If Teredo encapsulation is found and encapsulated IPv6 address is in format which is specified in [7], plug-in sets type of tunnel to indicate usage of Teredo and pass filled data structure to exporter. Detection of ISATAP and 6to4 packets is similar as they share some characteristics. IPv4 protocol must be set to value 41. In both mechanisms IPv4 header is followed by IPv6 header. To decide if IPv6 packet is encapsulated by ISATAP or 6to4 plug-in checks IPv6 addresses and looks for address in format specified for 6to4 or ISATAP. Filled data structure is passed to the FlowMon exporter.

4.4   Packet Processing Performance

Packet processing performance was measured as a throughput test when processing packets from 10 Gbps Ethernet link (see Figure 2). The measurement run on 2.0 GHz quad-core CPU and beside throughput we also monitored CPU usage (see Figure 3). Throughput was measured for Teredo and 6to4 packets (throughput of ISATAP packets is the same as throughput of 6to4 packets). In first scenario all packets were processed by single instance of the FlowMon exporter with loaded input plug-in. In the second scenario packets were distributed to 4 instances of the FlowMon exporter with loaded input plug-in. Each instance of the FlowMon exporter was running on different CPU core providing more computing power for processing.

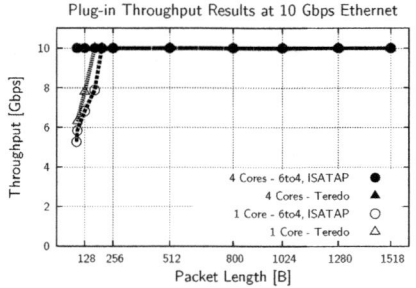

Fig. 2. Throughput on 1 and 4 CPU cores

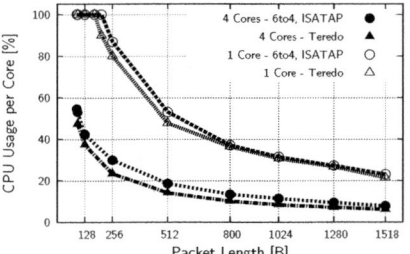

Fig. 3. CPU load on 1 and 4 cores

To minimize impact of flow generation on performance results all packets in the first scenario originated from single flow. In case of the second scenario four different flows were used.

5 Monitoring of Real Network

We deployed monitoring system based on the proposed approach on the CESNET2 network. Three 10 Gbps backbone links which are connecting the CESNET2 network to SANET (Slovak academic network), PIONIER (Polish optical Internet) and NIX.CZ (Neutral Internet eXchange of not only Czech Republic) networks were monitored. We were forced to slightly change the IPFIX templates in way they shouldn't be according to the IPFIX standard as we were using modified NfSen. NfSen doesn't have full support for enterprise-specific elements [13]. The presented statistics are from September 24 to October 6, 2010.

5.1 IPv6 Address Assignment

Address assignment is a little bit different in IPv6 networks. Usually stateless auto-configuration is used [11], so a host learns just network prefix and default gateway. The lower part of IPv6 address (last 64 bits) is a host identifier and can be assigned manually, based on EUI-64 algorithm or generated randomly according [2]. We use similar methodology for address classification as in [5] but some addresses are analyzed in detail.

Table 1. IPv6 unique addresses – average per day

Traffic	Unique Addresses	Note
Native IPv6	8059 (10.1%)	details in Table 3
6to4	20090 (25.3%)	details in Table 2
Teredo	51330 (64.5%)	detected 13 Teredo servers
ISATAP	82 (0.1%)	

Table 2. 6to4 addresses in detail

	Native	Tunneled Traffic
Autoconf	2.7%	1.4%
Linux	1.2%	0.3%
Windows	91.2%	85.6%
Privacy	4.9%	12.7%

Table 3. Global IPv6 addresses in detail

	Native	Tunneled Traffic
Autoconf	9%	4.2%
Privacy	69.2%	69%
Low	21.8%	26.8%

Table 1 shows average number of unique IPv6 addresses in native, 6to4, Teredo and ISATAP traffic per day. Surprisingly there is very high number of Teredo addresses. Further examinations showed that Teredo is used mainly for p2p sharing. We believe that it is because BitTorrent clients such as $\mu Torrent$ have implemented Teredo support, to be able to share data with more peers. We detected several Teredo servers as well.

Native and 6to4 addresses are more analyzed and results are shown in Table 2 and Table 3. First table describes in detail 6to4 addresses in native and tunneled traffic. Autoconf means, that EUI is generated according to EUI-64. Linux and Windows rows describes, how many hosts use Windows and Linux/Unix operating systems. This detection is based on default values for the EUI fields [12]. Privacy means, that EUI is generated according to Privacy Extensions. The second table shows address structure of global IPv6 address in native and tunneled traffic.

5.2 Tunneled Traffic Characteristics

The first interesting fact about IPv6 tunneled traffic is, according to our measurement, that it generates more traffic then native IPv6 traffic. This fact is true for all of three metrics (by flow, by packets and by bytes) and is shown in Table 4. As described earlier, the reason for this can be presence of tunneling mechanisms in recent versions of MS Windows.

Majority of IPv6 tunneled traffic uses Teredo mechanism (see Table 5). The least used mechanism is ISATAP that may be given by the fact that it is the least preferred option of tunneling in MS Windows.

Table 4. Traffic distribution

	Flows	Packets	Bytes
IPv4	98.39%	99.19%	99.13%
Native IPv6	0.10%	0.12%	0.21%
Tunneled IPv6	1.50%	0.69%	0.66%

Table 5. Tunnel distribution

	Flows	Packets	Bytes
Teredo	88.18%	89.10%	88.85%
ISATAP	0.06%	0.03%	0.03%
6to4	11.76%	11.76%	11.12%

We also observed very different distribution of application protocols in tunneled IPv6 traffic. The most used protocol in IPv4 and IPv6 traffic is HTTP. In tunneled IPv6 traffic its share was very small and the traffic was overall spread to hundreds of UDP and TCP ports with high numbers. We come to conclusion that tunneled IPv6 especially Teredo is used for p2p sharing. Reasons, why p2p programs use Teredo are described in Section 5.1.

Table 6. Protocol distribution in tunneled and native traffic

	Flows			Packets			Bytes		
	IPv4	IPv6	Tunnel	IPv4	IPv6	Tunnel	IPv4	IPv6	Tunnel
HTTP	38.25%	1.99%	0.35%	49.99%	65.50%	2.98%	56.80%	76.16%	0.38%
HTTPS	3.26%	<0.01%	0.08%	1.72%	<0.01%	2.85%	1.17%	<0.01%	0.33%
DNS	10.39%	61.76%	0.45%	0.45%	1.68%	0.05%	0.07%	0.42%	0.01%

6 Conclusion

Current flow-based traffic monitoring techniques can not easily analyze tunneled traffic. It is especially problem in IPv6 networks. In IPv6 networks tunnels are created automatically, without users or administrators intervention. Because IPv6 protocol is not compatible with current IPv4, these tunneling mechanisms would be needed for several years. Network administrators will need an approach, which is able to monitor tunneled traffic on high-speed networks, is scalable and can be integrated into current monitoring systems. In this paper we propose such an approach.

Monitoring 10 Gbps links is possible using hardware-accelerated network cards. We implemented a plug-in for the FlowMon exporter, which can monitor tunneled IPv6 traffic and export obtained data using IPFIX. Collected data can be further analyzed by IDS (Intrusion Detection System) and IPS (Intrusion Prevention System). Current monitoring software miss information about the tunneled traffic. We propose an approach which is able to monitor this kind of traffic. We successfully deployed the proposed solution on academic backbone links in the Czech Republic.

Acknowledgments. This work was supported by the Research Intent of the Czech Ministry of Education MSM6383917201 and by the Czech Ministry of Defence – contract no. SMO02008PR980-OVMASUN200801.

References

1. INVEA-TECH a.s., FlowMon exporter, cited [30.09.2010], http://www.invea-tech.com/products-and-services/flowmon/flowmon-probes
2. Narten, T., Draves, R., Krishnan, S.: RFC 4941, Privacy Extensions for Stateless Address Autoconfiguration in IPv6 (September 2007)
3. Shen, W., Chen, Y., Zhang, Q., et al.: Observations of IPv6 traffic. In: CCCM, vol. 2, pp. 278–282 (2009) ISBN 978-1-4244-4247-8
4. Karpilovsky, E., Gerber, A., Pei, D., Rexford, J., Shaikh, A.: Quantifying the Extent of IPv6 Deployment. In: Moon, S.B., Teixeira, R., Uhlig, S. (eds.) PAM 2009. LNCS, vol. 5448, pp. 13–22. Springer, Heidelberg (2009)
5. Malone, D.: Observations of IPv6 addresses. In: Claypool, M., Uhlig, S. (eds.) PAM 2008. LNCS, vol. 4979, pp. 21–30. Springer, Heidelberg (2008)
6. Savola, P.: Observations of IPv6 Traffic on a 6to4 Relay. ACM SIGCOMM CCR 35(1), 23–28 (2005)
7. Huitema, C.: Teredo: Tunneling IPv6 over UDP through Network Address Translations (NATs). RFC 4380 (February 2006)
8. Carpenter, B., Moore, K.: Connection of IPv6 Domains via IPv4 Clouds. RFC 3056
9. Templin, D.T.F., Gleeson, T.: Intra-Site Automatic Tunnel Addressing Protocol (ISATAP). RFC 5214 (March 2008)
10. Nordmark, E., Gilligan, R.: Basic Transition Mechanisms for IPv6 Hosts and Routers. RFC4213 (October 2005)
11. Thomson, S., Narten, T., Jinmei, T.: IPv6 Stateless Address Autoconfiguration. RFC4862 (September 2007)
12. Warfield, H.M.: Security Implication of IPv6, Internet Security Systems (2003)
13. Krejčí, R.: Network Traffic Collection with IPFIX Protocol, cited [2010-10-04], http://is.muni.cz/th/98863/fi_m/xkrejc14_dp.pdf

Identifying Skype Traffic in a Large-Scale Flow Data Repository

Brian Trammell[1], Elisa Boschi[1], Gregorio Procissi[2], Christian Callegari[2], Peter Dorfinger[3], and Dominik Schatzmann[1]

[1] ETH Zürich, Zürich, Switzerland
[2] Università di Pisa, Pisa, Italy
[3] Salzburg Research, Salzburg, Austria

Abstract. We present a novel method for identifying Skype clients and supernodes on a network using only flow data, based upon the detection of certain Skype control traffic. Flow-level identification allows long-term retrospective studies of Skype traffic as well as studies of Skype traffic on much larger scale networks than existing packet-based approaches. We use this method to identify Skype hosts and connection events to the network in a historical flow data set containing 182 full days of data over the six years from 2004 to 2009, in order to explore the evolution of the Skype network in general and a large observed portion thereof in particular. This represents, to the best of our knowledge, the first long-term retrospective analysis of the behavior of the Skype network based solely on flow data, and the first successful application of a Skype detection algorithm to flow data collected from a production network.

1 Introduction

In the last few years the Internet conferencing and instant messaging application Skype has become a key method of communication among users on the Internet. The proprietary nature of its algorithms and protocols and its extensive use of encryption make Skype traffic identification a challenging task. Traffic identification serves two broad purposes: the identification of nodes using Skype, and long-term retrospective analysis of the network-level behavior of a useful application whose protocol specifications are unpublished.

Unlike other VoIP applications, Skype uses a peer-to-peer overlay network for communication among Skype clients, and uses "supernodes" for message relaying and handling metadata such as user profile and presence information. Skype clients that are accessible from the open Internet (i.e., not NATted or firewalled) with adequate bandwidth and resources may be promoted to supernode status. Thus, Skype does not own most of the infrastructure of its network, itself providing only a relatively small set of "bootstrap" supernodes, gateways to the public switched telephone network, and a login server for identity management. Bootstrap supernodes are contacted only when a client is newly installed, and even the login server does not need to be contacted at each login, as clients may

J. Domingo-Pascual, Y. Shavitt, and S. Uhlig (Eds.): TMA 2011, LNCS 6613, pp. 72–85, 2011.
© Springer-Verlag Berlin Heidelberg 2011

use a supernode to relay login information. This architecture has allowed Skype to rapidly grow its network while minimizing its need to build out infrastructure.

Research on Skype has to date focused on techniques working with packet-level data, including payload. A *flow*, however, typically represents a set of packets sharing the same IP addresses, ports, and protocol; or one side of an exchange between two IP sockets. Flow data contains only this flow key, timestamps, and byte and packet counters. This represents a significant data reduction over packet traces for the same traffic. For many large-scale networks flow data is all that is available, or practical to collect.

In this paper, we build on existing work in packet-level Skype traffic classification to develop a novel method for determining the presence of Skype clients and supernodes on a network using only flow data, based upon the detection of certain Skype control traffic. We then apply this approach to the historical study of Skype traffic over six years in an existing flow data set collected from a medium-sized national-scale network. The use of flow data enables an examination of a much larger observed portion of the Skype network than in previous works.

The remainder of this paper is organized as follows: we first review related works in section 2. In section 3 we elaborate on the specific features of Skype signaling traffic that lend themselves to flow-based analysis, then provide details of the algorithm which enables that analysis. Section 4 evaluates our approach, against a proxy identification method for Skype traffic in flow data, active detection of Skype nodes, and an existing packet-based Skype detector. We then apply our algorithm in Section 5 to the examination of long-term trends in a very large network dataset covering the six years from 2004 to 2009.

2   Related Work

Research on Skype traffic classification and characterization has been an area of interest for some time, but has to date been focused largely on packet-level traffic measurements. Initial work focused on reverse engineering. In [3], the authors give a detailed overview of Skype architecture and functionality; and in [4] the authors examine the application from a network administrator's viewpoint. Control traffic classification at the packet level is detailed in [8] over a five-month period on a small network. Full classification follows in [6], which presents a real-time framework for Skype traffic detection based on two complementary techniques. The first technique is based on pattern recognition, looking for Skype fingerprints in the packet structure. The second leverages packet arrival rate and packet length statistics to feed a decision mechanism based on naive Bayesian classification.

More recently, Rossi *et al.* [13] study Skype signaling mechanisms through passive measurements and provide insights on the complexity of managing the Skype peer–to–peer overlay network. Bonfiglio *et al.* [5] characterize Skype traffic using both passive and active measurements. The authors use their classification to investigate user behavior, as well.

Active approaches are more suited to differential provisioning of Skype services on operational networks. Bremler-Barr *et al.* [7] "harvest" supernodes by preventing connection to them in order to enable Skype traffic filtering. Using both experimental results and an analytic model, the authors show that it is possible to collect enough supernode addresses so as to block the service for an arbitrary connecting client.

The present work, however, is most directly inspired by Adami *et al.* [1], who present a novel real–time algorithm for Skype traffic detection and classification based on the combination of signature matching and a statistical approach. The algorithm can recognize and classify some specific signaling message exchanges; we use these features as they prove to exhibit easily recognized signatures in flow data as well.

There have been prior attempts at flow-based Skype detection. Angevine and Zincir-Heywood [2] use augmented flow data generated from packet data as an input to an existing machine learning system for the purpose of detecting Skype traffic, However, we note this mechanism requires three particular flow properties, TCP acknowledgment count and minimum and maximum flow packet lengths, which are not supported by commonly deployed flow generators. Therefore, this flow-based method practically requires access to original packet data, and is not applicable to analysis on existing long-term, large-scale flow data repositories.

3   Methodology

Our methodology draws on some of the same features of Skype traffic identified in [1] to IP flow measurements based upon a detailed analysis of that algorithm and the visibility of the artifacts it measures in flow data. Here we will review the subset of features of the protocol we use for flow-level detection.

Skype uses a hybrid P2P overlay network of clients and supernodes. Supernodes are specialized client nodes which distribute directory and presence information, and relay messages on behalf of the clients. They can be thought of as the internal nodes of the network, with the clients being the leaf nodes.

Here we focus on two specific message exchanges between clients and supernodes which we call the *UDP probe* and *TCP handshake*.

3.1   UDP Probe and TCP Handshake

The UDP probe consists of a set of messages exchanged to discover a supernode with which to connect to the network, and the characteristics of the connection between the client and supernode (e.g., the presence of NATs, firewalls, etc.). The Skype UDP probe has two forms, a long probe shown in figure 1(a)(1) consisting of two packets in each direction, and a short probe shown in figure 1(a)(2) consisting of one packet in each direction. Either a long probe or a short probe may be seen in connection initiation.

After the UDP probe has been completed, the client initiates a TCP handshake if the size of the UDP payload of the last packet $y = 18$ bytes; we interpret

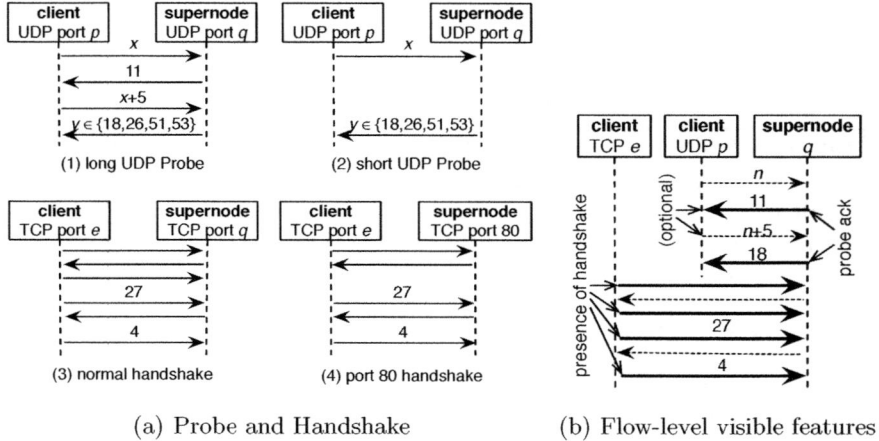

Fig. 1. Packets in Skype connection interactions. Numbers represent packet payload sizes in bytes, which are variable when not shown. Highlighted packets indicate flows used by our approach.

these messages as supernode acknowledgements. If, however, $y \in \{26, 51, 53\}$ bytes, the supernode will not be further contacted by the client in this session; we interpret these messages as negative acknowledgments. The UDP probe is repeated until a supernode is selected. Furthermore, the client periodically repeats the short probe, to ensure it always has an available supernode.

The fact that each packet sent from the supernode back to the client during a UDP probe has a known payload size makes this message exchange useful for detection at the flow level.

Concurrently with UDP probes at network connection time, the client will attempt TCP handshakes with any supernode that positively acknowledges the probe to TCP port q, which is the same port number to which the UDP probe was sent. The TCP handshake exchange is shown in figure 1(a)(3). If the TCP handshake connection cannot be opened, the client will instead use port 80 as shown in figure 1(a)(4), or fall back to port 443. These ports are selected as they are less likely to be blocked; we note specifically that the Port 80 exchange does not use HTTP.

3.2   Flow Level Detection

We illustrate the features of this exchange that are useful for flow-level Skype detection in figure 1(b). Both the short and long UDP probes have easily recognizable signatures in flow data, and are followed relatively rapidly by the handshake, which is attempted on one of three predictable TCP ports $q, 80, 443$. Specifically, the downstream side of a long probe consists of a 2-packet UDP flow with 85 total bytes, while the downstream side of a short probe consists of a 1-packet UDP flow with 46 total bytes. We derive these flow sizes by adding the size of the IP and UDP header on each packet to the sizes of the packet payloads in the UDP probe.[1] Simply searching for this pattern in flow traffic as shown in algorithm 1 is then sufficient to recognize Skype supernodes and clients given a traffic stream with a high degree of fidelity.

Algorithm 1. Recognition of supernodes and clients given a set of flows

---

$acklist \Leftarrow \emptyset$, $supernodes \Leftarrow \emptyset$, $clients \Leftarrow \emptyset$
for all $f \in flowstream$ **do**
  if $protocol(f) = $ UDP and $port_{dest}(f) \geq 1024$ **then**
    if $\langle packets(f), bytes(f) \rangle \in \{\langle 1, 46 \rangle, \langle 2, 85 \rangle\}$ **then**
      $sn \Leftarrow address_{source}(f)$, $q \Leftarrow port_{source}(f)$, $cl \Leftarrow address_{dest}(f)$
      $acklist \Leftarrow acklist \cup \langle sn, \{q, 80, 443\}, cl \rangle$
    end if
  else if $protocol(f) = $ TCP and
       $packets(f) \geq 3$ and
       PSH $\in flags(f)$ *(if flags present)* **then**
  $sn \Leftarrow address_{dest}(f)$, $q \Leftarrow port_{dest}(f)$, $cl \Leftarrow address_{source}(f)$
  if $\langle sn, sp, cl \rangle \in acklist$ **then**
    $supernodes \Leftarrow supernodes \cup sn$
    $clients \Leftarrow clients \cup cl$
  end if
  end if
end for

In this approach we take two empirically-determined measures to reduce false positives. First, we reject UDP traffic on well-known ports, as Skype does not in the general case use ports lower than 1024 for supernode acknowledgement, and because the pattern detected by this algorithm on port 53 is consistent with a TCP answer to a UDP DNS query. Second, the *acklist* in algorithm 1 must also be periodically expired; in our implementation, all *acklist* entries are guaranteed a minimum lifetime of one second, as experimentation showed that the vast majority of successful acks are answered by handshakes within one second.

Algorithm 1 recognizes client connection events at the flow level, enabling the study of Skype traffic in data sets containing only flows[2]. It does not attempt general traffic classification. It specifically ignores the data plane, or signaling other than that at connection time.

The algorithm does have a few important limitations. Relying as it does on UDP and TCP, it is incapable of identifying Skype client connections from networks where UDP is blocked or disabled at the client [14], and the client uses an alternate handshake method; this, we suspect, is a significant component of the

---

[1] The presence of IP options may increase the size of each of these flows in 4-byte increments; however, in our initial evaluation of this algorithm on several days of flow data, enabling IP option detection had no impact on the results; all handshakes were seen after 46- or 85-byte flows.

[2] Note that the examined data set is missing TCP flag data, so our evaluations do not make use of the PSH flag check in this algorithm. Including TCP flag data may increase detection fidelity, but we have not evaluated this case.

real false negative rate, as we elaborate in section 4. It relies as well either on flow traffic captured in both directions, or additional TCP flag information in the supernode-to-client direction that would allow the identification of a successful client-initiated TCP handshake flow. It is not adaptable to sampled flow data, or flow data assembled from sampled packets, because it observes multi-flow interactions and requires specific flow packet count and size data.

The development of any detection algorithm must be concerned with the possibility of evasion. In this case, the interactions we detect are fairly basic to the operation of Skype protocol, so the properties of these interactions themselves would need to be redefined in order to evade this detector. How easy this would be, in a backward-compatible and realistically deployable way, is unknown. However, this possibility should be taken into account in any effort to adapt this algorithm for online Skype client detection, for example, for differentiated services purposes.

4   Evaluation

We implemented this algorithm in a detector, which we dubbed snack because it tracks supernode (SN) positive acknowledgments (ACK) of UDP probes. We then set about measuring its accuracy and performance. First, we ran a test protocol containing two distinct clients, one of which starts four times, amid web surfing, secure shell, and VPN background traffic; both clients and all five connection events were correctly identified in each of three different trials, all within twenty seconds of the start of the Skype application, without any false positives.

We then sought to evaluate our algorithm against data collected from larger networks. Here we have a problem with selecting a good proxy for ground truth. We use three different evaluation methods, each with its own advantages and disadvantages: passive comparison against traffic to a host known to be used by the Skype client, verification using active Skype service discovery, and comparison to an existing packet-based detector. In all three of these, we focus on the ability of snack to detect clients. The summary of these results is shown in Table 1.

4.1   About the Observed Network

The flow data used in evaluation and exploration in this work comes from a flow data set from the border of SWITCH [15], the Swiss national research and education network. SWITCH operates a production network providing connectivity to the Internet for universities and research laboratories across Switzerland.

This network advertises prefixes for about 2.31 million IPv4 addresses, and the typical daily traffic volume is between 50 and 100 terabytes. The data set is made up of hourly data files containing on the order of 200 megabytes to two gigabytes of compressed flow data per hour. We studied one small portion of this dataset representing four full days in February 2009 for the evaluation against

Table 1. Results of snack evaluation on the measured network in sections 4.2, and 4.3, and 4.4

update server eval (4.2)	$clients$	$updaters$	$c \cap u$	FP_{max}	FN_{max}
Su 15 Feb 2009 (24h)	1846	1811	1224	34%	32%
Mo 16 Feb 2009 (24h)	7913	7116	5648	29%	21%
Tu 17 Feb 2009 (24h)	8087	7457	5798	28%	22%
We 18 Feb 2009 (24h)	7769	7386	5513	29%	25%
$mean$				30%	25%
$union$ (96h)	13323	11809	9823	26%	17%

active nmap eval (4.3)	$clients$		$verified$	FP_{max}	
Th 6 May 2010 (5.0h)	4008		3618	10%	

packet-based eval (4.4)	$clients_{pkt}$	$clients_{flow}$	$c_p \cap c_f$	$size_{pkt}$	$size_{flow}$
Tu 6 Oct 2009 (7.5h)	17	13	13	11000	41
Tu 13 Oct 2009 (10.3h)	17	15	15	48000	100
Fr 30 Oct 2009 (34.3h)	18	16	16	46000	165

the update server in the section 4.2, one small portion of this data on one day in May 2010 for the evaluation against active probing in the section 4.3, and one larger portion of the dataset representing 182 full days[3] of traffic in both directions across the network border for the entire month of August for each year from 2004 to 2009, inclusive, in section 5 to examine long-term traffic trends in the observed portion of the Skype network.

During the period under observation in the larger dataset, the size of the network was more or less constant in terms of routable IP space, fluctuating less than 9%. It grew from about 2.28 million IPv4 addresses in 2004 to 2.48 million addresses in 2006, falling slowly to 2.31 million addresses by 2009.

The measured network is an access and interconnection network for research institutes and universities without a significant residential population, so it is somewhat biased toward weekday, working-hour traffic. The total flow volume exhibits a characteristic two-peak daily seasonality, with peaks around 07:00 and 12:00 UTC (09:00 and 14:00 local time), corresponding to daily activity cycles of the connected users; we will see this pattern emerge in Skype traffic as well.

4.2   Evaluation against Update Server Activity

Skype clients periodically query an update server[4] to determine whether a newer Skype client is available for download. We assume in this evaluation that this host is used only by Skype clients. We can therefore use the presence of traffic to this server in the data set to evaluate worst-case false positive and false negative rates.

We estimate an upper bound on the false negative client detection rate by counting any host internal to the network contacting the update server within

---

[3] 2005 excludes four days of data due to a measurement system outage.

[4] As of February 2009, the update server was ui.skype.com (204.9.163.158).

a given day, but not detected as a client, as a false negative. Note that this is only an upper bound for two important reasons. First, there is no temporal correlation between the connection events detected by snack and update server contact, and second, snack cannot detect internal clients contacting internal supernodes. However, we do estimate that a non-trivial component of this false negative rate is real, corresponding to nonmeasurement because UDP traffic is blocked or disabled at the client.

We can also provide an upper bound on the false positive client detection rate by counting any host inside the network detected as a client, but not contacting the update server within a given day, as a false positive. Similar to the case above, this is only an upper bound, as the Skype application allows the user to disable checking with the update server [14].

Here we consider each day separately and take the mean of the false positive and false negative rates, yielding a mean maximum false positive rate of 30% and a mean maximum false negative rate of 25%. When considering the union of each address set over the four days, these rates go down to 26% and 17% respectively. This reflects the general lack of temporal correlation between connection and update.

4.3   Bounding False Positives: Active Supernode Identification

The widely-used network scanning tool nmap supports active Skype detection, as described in [11]. In this evaluation, we took addresses of Skype nodes from snack fed these to nmap for verification. This evaluation has two important limitations. First, it can only provide a false positive rate (detected by snack but not nmap). Second, due to delays inherent in the data collection and distribution system, a maximum of four hours pass between the snack Skype detection and the nmap probe. We presume that supernodes are relatively more stable than the average host, and will have longer uptimes, which should minimize the impact of this delay. We attempt to control for host shutdown by ensuring the host responds to ICMP echo requests (pings), but this method is itself imperfect: first, not every host will respond to pings, or may be behind an ICMP-blocking firewall. Second, this method cannot control for application shutdown or dynamic addressing changes. Therefore, we can only provide an upper bound on false positive rate as in section 4.2.

We ran this evaluation over five hours during the workday on May 6, 2010. We fed a sample of 14149 detected Skype nodes to nmap. Of these, 4008 were up at the time of the nmap probe, 3618 of which were positively identified as Skype nodes. This translates to a maximum false positive rate of less than 10%, significantly better than that indicated by the evaluation based on update server contact.

4.4   Comparison to Existing Packet-Based Approach

In this subsection, we compare the performance of snack to that of the packet-based detector developed by Adami et al. [1]. This comparison was done on packet data collected during October 2009 from a small network, in this case a single 100Mbit link of a small local cable provider which is not part of the network described in section 4.1. This link provides Internet access to about 30 small businesses and about 70 residential users representing a range of different local network topologies (e.g. NATted, firewalled, and directly connected client machines) and client operating systems. This makes it necessary for Skype to adapt its connection process to the various conditions.

We then generated flows from the packets using YAF [9]. As shown in the *size* solumns in table 1, the original packet traces were between about 200 and 500 times larger than resulting flow files, illustrating the data reduction typical of flow-based processing.

The flow files were then given to snack, and the packet files to the Skype detector detailed in [1]. Here, note that if we treat the packet-based detector as ground truth, on this network snack has a false positive rate of zero; i.e., every client detected by snack is also detected by the packed-based detector. The false negative rate is between 11% and 23%, within the bounds but lower than the maximum false negative rate detected in section 4.2. Manual inspection shows that at least half the difference between the packet-based detector and snack can be accounted for due to the inability of snack to detect connection events in the absence of UDP traffic during association.

We note that, even factoring in the time required to generate flows from the packet trace, snack significantly outperforms the packet-based approach, requiring about seven minutes (434s) to find clients in the first (7.5h on Tuesday 6 October) trace, as opposed to three and a half hours (12423s), for a speedup factor of about 30.

4.5   A Note on Performance

snack is quite lightweight, and intended to be integrated into existing large-scale flow processing workflows for retrospective and on-line analysis. The analysis for this study could easily be run in "real time" on a national scale network: as an example, during the run of the study in section 5, snack processed 101 gigabytes of compressed flow data in export order[5], covering 10800 minutes (one full week) and containing 230793 connection events, in 670 minutes.

However, the key performance benefit of flow-based analysis is the data reduction achievable with flow data. The data set described in section 4.1 would require on the order of 10 terabytes, 100 terabytes, or one petabyte of storage per month to support analyses on packet header, partial packets (128 bytes of payload) or full packets, respectively. Long-term full packet storage of this magnitude is neither technically nor legally feasible. Indeed, it is the multiple-orders-of-magnitude improvement over packet data in storage efficiency and the reduced privacy impact that makes large-scale network studies and operational measurement such as this one practical, and the primary reason we sought a flow-based method for Skype detection.

---

[5] Our flow data set is stored as exported by the router; therefore, performance figures here include the time required to reorder the flows by end time as required by snack.

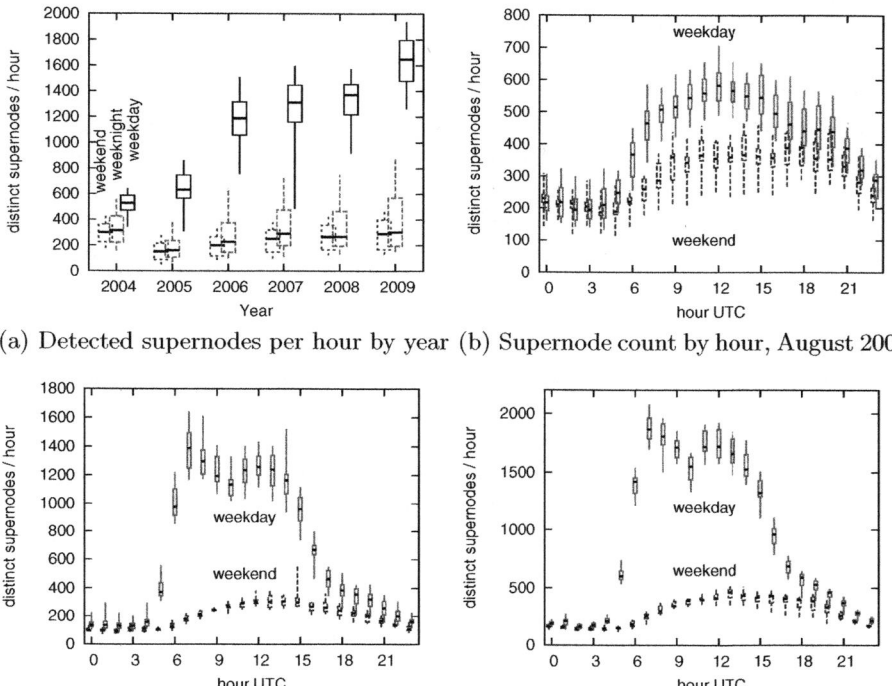

(a) Detected supernodes per hour by year (b) Supernode count by hour, August 2004

(c) Supernode count by hour, August 2006 (d) Supernode count by hour, August 2009

Fig. 2. Growth of the Skype network

5   Insights on the Skype Network

As noted in section 4.1, we ran `snack` over a flow data set from the border of SWITCH. We explored trends over time in the development of the observed portion of the Skype network, using the number of distinct supernodes seen per hour as a proxy measurement for the size of the Skype network.

This measurement shows strong weekly and daily seasonality, which is to be expected given that the number of observed supernodes correlates to human communication activity; therefore, in the remainder of this section we will examine quartiles[6] of hourly data sets as opposed to raw time series.

5.1   Network Size

In figure 2(a), we show the observed supernode counts for weekends, weeknights (16:00-05:59 UTC), and weekdays (06:00-15:59 UTC) in August in each of the measured years[7]. The median observed daytime network size increases rapidly

---

[6] Quartile plots in this section show the 5th, 25th, 50th, 75th, and 95th percentile measurement of each examined variable.

[7] 2007 data excludes data during the 16 August Skype outage. See [12] for more.

to 2006 as Skype grows more popular, then continues increasing more slowly thereafter, with another smaller peak in 2009.

The increase in the size of the Skype network during the day well outpaces the increase in the size of the SWITCH network mentioned in section 4.1.

In comparison, observed night and weekend network size remain relatively constant. This may derive from the relatively light weekend and overnight traffic on the observed network, but we nevertheless interpret this finding to mean that the observed portion of the network has a minimum base size; as clients disconnect from the network at the end of the day, the Skype network maintains a certain number of supernodes. The network then adapts above this base to cope with client load.

In more detail, from 2004 (in figure 2(b)) to 2006 (in figure 2(c)), the median peak hourly size of the network increases from about 600 to about 1400 distinct supernodes; then to about 1850 through the stabilization phase in 2009 (in figure 2(d)). From 2006, the daily pattern shows two distinct peaks at 07:00 and 12:00 UTC (09:00 and 14:00 UTC+2, Central European Summer Time, which is the local time on the observed network). As noted in section 4.1, this pattern is a characteristic of all traffic on this particular network, and we interpret it to represent two distinct peaks in human activity split by a mid-day break.

5.2   In-Degree of Supernodes

We then examined the number of distinct clients connecting per supernode per hour, which allows us to estimate the "in-degree" of supernodes. For in-degree measurement, we focus only on supernodes internal to the network[8], as we assume that the fact that the external network is much larger than the internal network implies that the number of clients per supernode is approximately as measured.

Here, in figure 3(a), we see a long-term downward trend from a median in-degree of 15.1 in 2004 to 7.13 in 2007, raising slightly to 9.33 in 2009. Note too that the shape of the distribution changes after 2007, with the 75th and 95th percentile numbers rising from 10.3 to 18.0 and 18.4 to 25.3 respectively. This indicates better balancing of the client load among supernodes throughout the stabilization phase, as there are more higher-degree supernodes in later years than in earlier years.

We examine supernode degree in more detail by measuring per hour in August 2009, shown in figure 3(b). Note that at high-traffic times during the working day, the number of supernodes per client is relatively low, with a median between six and seven. As clients and supernodes begin shutting down at the end of the day, external clients must make do with fewer available supernodes; the number of clients per supernode then peaks in the evening, with fewer supernodes (generally on well-connected office networks) serving more home users (generally on less-well-connected residential networks). As with the size of the observed network, weekend and overnight traffic are similar; on weekends, the median number of clients per supernode hovers around twenty.

---

[8] Approximately one fifth to one third of all supernodes observed, on an hourly basis, are internal to the network.

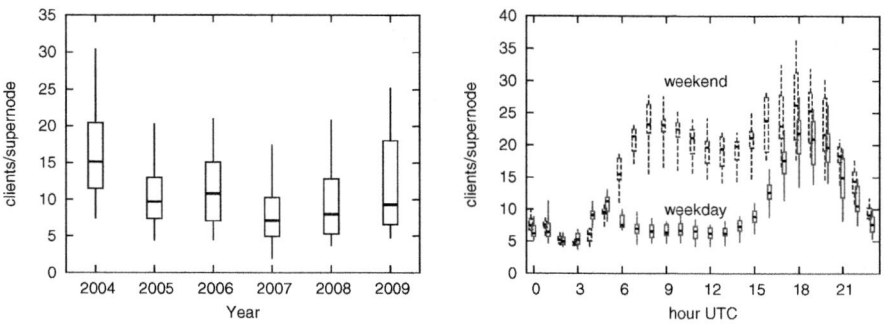

(a) External clients per supernode per (b) Hourly clients per supernode, August
hour   2009

Fig. 3. Supernode in-degree

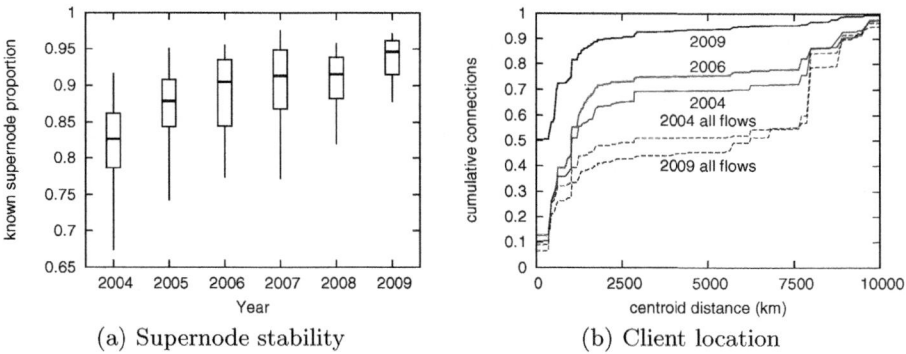

(a) Supernode stability   (b) Client location

Fig. 4. Trends in Skype network maturity

5.3 Network Maturity

In order to further explore the stabilization of the network, we searched for
metrics that can be used to estimate the stability of the observed portion of
the Skype network. Here we can use the proportion of connections detected by
snack on a given day involving a supernode that has already been seen that
day as a proxy for supernode stability. Here we see a long-term upward trend
in figure 4(a). On a median day in 2009, 94.6% of client connections are to a
known supernode, as opposed to 82.6% in 2004.

Another proxy for network maturity is its geographic performance. To a first
approximation, network distance is related to geographic distance; an overlay
network such as Skype is therefore performing well when the traffic of the observed portion of the network is biased toward the locality in which it is measured. In figure 4(b) we show the cumulative geographic distribution of clients
external to the observed network, i.e. those which contact supernodes within the
network, by the country code associated to IPv4 address by the MaxMind GeoIP
database [10], in kilometers distant of each country centroid from the observed

network in 2004, 2006, and 2009. Indeed, here we see a marked increase in locality between 2006 and 2009. Specifically, the proportion of client connections within 2000 kilometers of the observed network, roughly corresponding to the European continent, rises from 63.4% in 2004 to 71.6% in 2006 to 89.6% in 2009. Note also that two large steps in the 2004 and 2006 distributions around 7500 kilometers, corresponding to large numbers of clients in China and the United States, are not present in the 2009 distribution.

We also compare each client distribution with the distribution of external IP addresses for all complete TCP flows on a typical day in 2004 and again in 2009. The proportion of external flows within 2000 kilometers is only 48.2% in 2004, falling to 40.9% in 2009. Skype client connections, therefore, demonstrate considerably better locality than all traffic in general, even as early as 2004, and even as the background traffic becomes more geographically diffuse.

6   Conclusions and Future Work

In this work, we have developed and evaluated snack, a flow-level Skype peer detector, by adapting existing work on packet-level reverse engineering of the Skype protocol to flow analysis. This represents, to the best of our knowledge, the first Skype peer detector that operates solely on flow-level data. Significantly, this approach scales to much larger networks than packet-based approaches. We then leveraged this new system to develop insights on the Skype network over a six-year time period from retrospective analysis of a flow-data archive collected from the edge of a national-scale backbone network; this is also the largest such portion of the Skype overlay network to be passively observed in the literature.

We believe the general approach we used in this work demonstrates the impact that packet-level reverse engineering efforts can have on flow-level network traffic analysis, as well as the ability of flow-level analysis to extend the scalability and applicability of features observed at the packet level.

Obvious future directions for the algorithm developed in this work include its application to other similar large-scale flow data repositories. In applications where simply knowing a given address is a Skype client or supernode at a given point in time is enough, the algorithm can be applied directly for operational Skype traffic detection, extending the ability to detect Skype peers to those networks with only flow data available. In cases where more detailed Skype traffic characterization is necessary, the identification of peers provided by our algorithm can be used as a "hint" to a second stage analysis; building such a multistage traffic detector is another area for future work.

Acknowledgments

The authors thank the FP7 PRISM and DEMONS projects for their support of this work. We would like to acknowledge SWITCH, the Swiss Research and Education Network, for providing the data used in this study. Thanks to thank Teresa Pepe and Arno Wagner for their feedback and assistance with evaluation of the approach in this paper.

References

1. Adami, D., Callegari, C., Giordano, S., Pagano, M., Pepe, T.: A real-time algorithm for skype traffic detection and classification. In: Balandin, S., Moltchanov, D., Koucheryavy, Y. (eds.) ruSMART 2009. LNCS, vol. 5764, pp. 168–179. Springer, Heidelberg (2009)
2. Angevine, D., Zincir-Heywood, N.: A preliminary investigation of Skype traffic classification using a minimalist feature set. In: ARES 2008: Proceedings of the 2008 Third International Conference on Availability, Reliability and Security, pp. 1075–1079. IEEE Computer Society, Washington, DC (2008)
3. Baset, S.A., Schulzrinne, H.G.: An analysis of the Skype peer-to-peer internet telephony protocol. In: INFOCOM 2006, 25th IEEE International Conference on Computer Communications (April 2006)
4. Biondi, P., Desclaux, F.: Silver needle in the Skype. In: Black Hat Europe 2006 (March 2006)
5. Bonfiglio, D., Mellia, M., Meo, M., Rossi, D.: Detailed analysis of Skype traffic. IEEE Transaction on Multimedia 11(1), 117–127 (2009)
6. Bonfiglio, D., Mellia, M., Meo, M., Rossi, D., Tofanelli, P.: Revealing Skype traffic: when randomness plays with you. SIGCOMM Computer Communications Review 37(4), 37–48 (2007)
7. Bremler-Barr, A., Dekel, O., Levy, H.: Harvesting Skype super-nodes. In: OWASP (December 2007)
8. Guha, S., Daswani, N., Jain, R.: An experimental study of the Skype peer-to-peer VoIP system. In: IPTPS 2006: The 5th International Workshop on Peer-to-Peer Systems, Microsoft Research (2006)
9. Inacio, C., Trammell, B.: YAF – Yet Another Flowmeter. In: 24th USENIX Large Installation System Administration Conference, LISA 2010 (November 2010) (to appear)
10. MaxMind. Geoip address location technology, http://www.maxmind.com/app/ip-location
11. nmap.org. Version detection using nse, http://nmap.org/book/nse-vscan.html
12. Rossi, D., Mellia, M., Meo, M.: Evidences behind Skype outage. In: Proc. IEEE International Conference on Communications 2009 (June 2009)
13. Rossi, D., Mellia, M., Meo, M.: Understanding Skype signaling. Computer Networks 53(2), 130–140 (2009)
14. Skype. Guide for network administrators, http://www.skype.com/security/network-admin-guide-version2.2.pdf
15. SWITCH. The Swiss Education and Research Network, http://www.switch.ch

On the Stability of Skype Super Nodes

Anat Bremler-Barr[1] and Ran Goldschmidt[2]

[1] Interdisciplinary Center, Herzliya, Israel
bremler@idc.ac.il
[2] University of Haifa, Israel
ran.goldschmidt@gmail.com

Abstract. The heart of skype services, one of the most ubiquitous P2P networks, is based on a set of super nodes. Choosing stable SNs is an important task, since it improves the whole performance and quality of the P2P network [1, 2]. In this paper we shed light on the life cycle of SNs using extensive data sets on Skype Super nodes, which were gathered over a period of 3 months. We then suggest how to choose a more stable SNs set.

The dynamic of nodes is inherent to the use of a computer, which is unplugged for some time or is mobile. Hence it is natural to predict that a Super Node would have multiple sessions correlated with the time the computer is up. Surprisingly, we show that 40% of the Super Nodes have only one session, with median residual life time of 1.75 days. These nodes also have a significantly shorter lifespan than Super Nodes that have multiple sessions, which have median residual life time of 67.5 days. We propose and give evidence that nodes with one session are nodes with dynamic IP addresses, and hence they have ended their life cycle due to a change of IP address. We show that the nodes with multiple sessions are mostly nodes with static IPs, and that choosing super nodes with static IPs would increase the availability and stability of the P2P network significantly.

1   Introduction

In P2P network that uses Super Nodes (SNs), choosing stable SNs is an important task, since the super nodes serve as the control tier for all the P2P nodes. The dynamic nature of P2P networks, where there are consistent changes in the set of nodes that participate in the p2p service, poses a huge challenge: designing a reliable service, while the core of the service is based on dynamic set of super nodes. By choosing stable Super Nodes, the whole performance and quality of the P2P network can be improved [1, 2].

Our paper sheds new light on the dynamic nature of SNs. We have collected an extensive data set of SNs from the ubiquitous P2P network, Skype. In a short period of time of 15 minutes we have collected 10,000 active SNs and followed their life cycles over a a long period of 3 months. We show that the SNs enjoy longevity: median residual life of 22.7 days and with median session length of 3 days, where a session of a node is a consecutive time that a node

J. Domingo-Pascual, Y. Shavitt, and S. Uhlig (Eds.): TMA 2011, LNCS 6613, pp. 86–99, 2011.
© Springer-Verlag Berlin Heidelberg 2011

is up. Surprisingly, we show that high percentage (40%) of the SNs has only one session. Inspired by this fact, we classify the SNs into two groups according to the number of sessions they have during their life time: Single Session in Life Time (SSLT) SNs group and Multiple Sessions in Life Time (MSLT) SNs group. We measure different parameters (life time, session length and availability) on the two groups, and found out that the nodes in the MSLT are more stable: with median residual life length of 67.5 days, while the SNs in the SSLT have median residual life of 1.75 days. Moreover we show that if we choose nodes only from the group of MSLT the P2P system would be more stable and have a lower churn (less by 19%) and higher accessability (higher by factor 2.1). We then show that SSLT is primarily composed from dynamic IP addresses. A dynamic IP address, changes its IP address from time to time (usually after it disconnects from its ISP). From the P2P point of view, an SN that changes its IP no longer exists in the P2P network and it is considered as a new SN node. The MSLT is primarily composed from SNs with static IP addresses, which are usually servers or part of academic networks and due to their primarily job need to be up most of the time and hence they are more stable.

While the dynamic nature of P2P network was extensively researched in many papers [2–8] our paper reveals new findings due to several reasons: the research is on super nodes and not on the regular peer nodes; we check the stability of the SNs over long time period; we measure Skype and not file transfer applications. Those differences may explain the main reason why the role of dynamic IPs was left in the shade until now. Most of the previous papers concentrate on file transfer p2p applications such as Gnutella, BitTorrent and Kad where the dynamic nature of nodes is more likely due to user activities such as shutting down the application after completing the task of downloading the file. And indeed, measurements on these networks observe sessions with length of a couple of hours [8]. However, in Skype usually the application is on all the time the computer is on (see Section 3), hence the dynamic of nodes is due to a network event, such as disconnecting the computer from the network. And indeed we measure median session duration of a couple of days. Moreover we measure subsets of the skype clients, the Super Nodes, which have relatively long lifes and the effect of the dynamic IP address which occurs in long time scale is shown. Note that those nodes were nominated by Skype to be Super Nodes. Even though the code of the Skype application is confidential and unknown, it is reasonable to assume that Skype tries to nominate peers that are more stables to be super nodes. The only previous paper that concentrated on Skype SN measurements by Guha *et al.*[9] did not concentrate on the stability of SNs, moreover its data was not suitable for measuring and understanding stability. That paper [9] checked only Super-Nodes that were alive after a period of 3 months from the time they were collected. Note that using our measurement we discover that only 20-30% of the SNs are still alive after 3 months.

One important conclusion from our work is a simple guideline of how to choose Super-Nodes. The previous technique[7], focuses on algorithms that take as input the history of the Super-Nodes. We show, that a key impact of stability of the super-Nodes is its IP address type. Super-nodes from static IP addresses have greater chance of remaining alive for a long period of time in the network, and hence should be preferred. While classifying the type of address (static or dynamic) using the IP address alone is a hard task, this is an easy task to the P2P client application, which can observe all the outgoing traffic and can detect the constant change in IP addresses and thus can conclude the type of the IP address.

2   The Model

In this section we model our P2P network, a partially centralized architecture[10] (i.e., P2P with SNs). The model suits the Skype P2P system model, but also suits other P2P networks with super nodes such as Kazaa[11], joost[12], iMesh[13], Morpheus[14]. The model is also applicable to fully distributed P2P networks, where we can consider each client as an SN.

The participating nodes in our P2P system are divided into two categories: Super-Nodes (SNs) and ordinary clients. The Super-Nodes create the control level, and basically are regular clients with good network connections that the P2P network decided to nominate to be SNs. Clients request control services[1] only through the super nodes. Each client maintains a list that contains a subset of the SNs, and when it wishes to connect the P2P network it picks one of the SNs in its list. Roughly speaking if none of the SNs in the list are active then the client cannot connect to the service. This is not completely accurate, since there are usually also bootstrapping servers that are located in the premises of the company that handles the P2P service. However, the P2P cannot rely on these servers due to the fact that these known servers are vulnerable to filtering attempts (the incentive of filtering P2P traffic is discussed in Section 5). Moreover the scalability of the P2P network is based on the fact that most of the clients are connect to the SNs and not to the main servers. A Super-Node is defined according to its IP and Port. At any given time an SN is in one of the following states: up (available) or down (fail). Nodes fail and recover according to some unknown process. A *session* of an SN is the continuous period of time the node is up.

We assume that the client's SN list is continuously updated with the "up" SNs as long as the client is connected to the P2P network. In the next section we show that our experiments with Skype clients reveal that this is the case with Skype.

---

[1] Such as connections to the P2P network, querying the P2P network on the IP of the callee and so on. After finding the IP address of the callee the caller communicates directly with the callee and initiates the call. If the client is unable to communicate directly with another client, then the SNs can also relay all communication, thus effectively bypassing firewalls.

3   Experiment Methodology

In spite of its massive popularity, little is known about Skype's inner-workings. Skype is a closed-source application and consequently, Skype Ltd. does not disclose its protocols and architecture. Extensive studies [9, 10, 15–24] were conducted to disclose the Skype architecture, protocols and inner-workings by using reverse engineering and measurement techniques. The precise algorithm of how Skype chooses which clients to nominate to SN and which SNs would serve a specific client is unknown.

In this paper we use the fact that in Skype versions 2-2.5 each Skype client holds a list of up to 200 SNs and their connection ports in a specific XML file (%appdata%\Skype\shared.xml). This SN list is not encrypted in those versions, while in later versions Skype encrypts the SN list.

Our first goal is to measure the changes in the SN list of a regular client. In Figure 1 we measure the changes in the SN list over time. We took a snapshot of the list at time t_0 and then at each time unit t_i we calculate the percentage of SNs that did not change and appeared also in the original SN list at time t_0. We repeated the test on 100 Skype clients and for duration of 2000 minutes. Figure 1 (see line labeled "SN List Regular Updates") shows that the SN list is updated constantly and that after 2000 minutes, which is a little less than a day and half, only 64% of the original SNs still appear in the SN list. On average Skype client received an update every 20-30 minutes.

Motivated by the fact that the SN list is constantly changing, our next step is to measure the dynamics of SNs, specifically the on and off time of the SNs. We first collected a set of SNs from multiple clients at the same time. In this process, named by us *SN extraction*, we extract SNs at higher rate by repeatedly doing the following on a Skype client: 1. Extracting the SN addresses and ports from the XML file; 2. Flushing most of the SN addresses from the list - leaving only specific SNs 3. Restarting the Skype client and waiting until the list is refreshed with 200 SN addresses. Each such iteration, takes approximately 2-2.5 minutes. By implementing the described process at the same time on 20 computers, located at ETH and Israel, we gathered 10,000 SNs with unique IPs in less than 15 minutes.

The *SN extraction* process is a modification of the method used by Guha [9], designed to work at a faster rate. We estimate that we gathered a snapshot of around 20% of the active set of SNs since the common estimation is that there are around 45K active SNs at any give time [25]. Figure 1 (see line labeled "SN List Extraction") shows the percentage of SNs that also exist in the SN list at time t_i that appeared also in the original SN list at time t_0. After 2000 minutes, the percentage is negligible and is only 2.5%. Thus, the flushing phase at the *SN extraction* causes Skype to send almost entire new SN list.

In order to check the status of an SN (i.e., if the SN is up or down) we use a Skype application ping, i.e., we send to the SN the first UDP packet of the Skype login process (similar technique to [9]). We checked the 10,000 SNs, every 15 minutes, where in each such iteration, we pinged all the 10,000 SNs

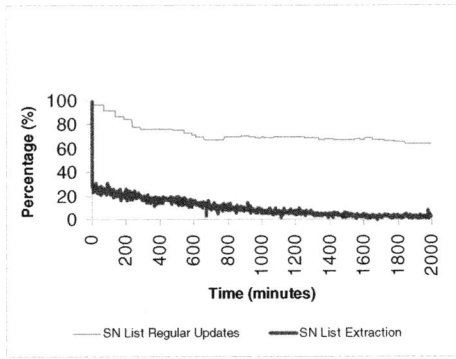

Fig. 1. The percentage of SNs at SN list at time t that appeared at the original SN list at time 0 in two scenarios: regular updates from Skype network and SN extraction

twice and waited for up to 5 minutes for the answer. An SN is considered to be up in iteration i if there was a response to at least one of the two pings in that iteration, otherwise the SN is considered down at iteration i. A *session* is defined as consecutive iterations where the SN is up. The SN life time is defined as the time elapsed between the start of the first session until the end of the last session.[2] We note that we verified that the Skype application ping is a good indicator of the fact that the client is still SN and is still alive. For dozens of SNs, we checked and verified that if the Skype application ping indicates that the SN is up, then we are able to connect to skype using this SN solely.[3]

Overall we did experiments over three months beginning on Apr 3. 2009. Our infrastructure was very stable and we conducted more than 9,000 iterations, and experienced connectivity problems only in 10 iterations. The stability and the large data set overcomes the known pitfalls in measuring the stability of SNs [8].

We note that there is a high correlation between the time the SN is up and the time the corresponding computer is up (and vise versa). Thus there is a good indication that usually the skype client is by default open at the client computer. We think that this is because the skype client is always on in order to be on standby to receive calls. The way we verified that a computer is on is by sending an ICMP ping to the computer. While most computers do not response at all to ICMP ping, since the ICMP is filtered in the host network for security reasons, a small subset of the SNs (17%) do respond to ICMP. We concentrated on this group and found that in those computers there is a decisive correlation between the responsiveness to the Skype application ping to the responsiveness to a regular ICMP ping (and vise versa).

---

[2] In order to avoid noises due to packet loss, we decided that a session ended only if at least in three consecutive iterations the SN was down. We found out that the results are not sensitive to this threshold.

[3] This was done by modifying the SN list at a client to hold only the examined SN and by filtering into the firewall any default hard coded skype servers.

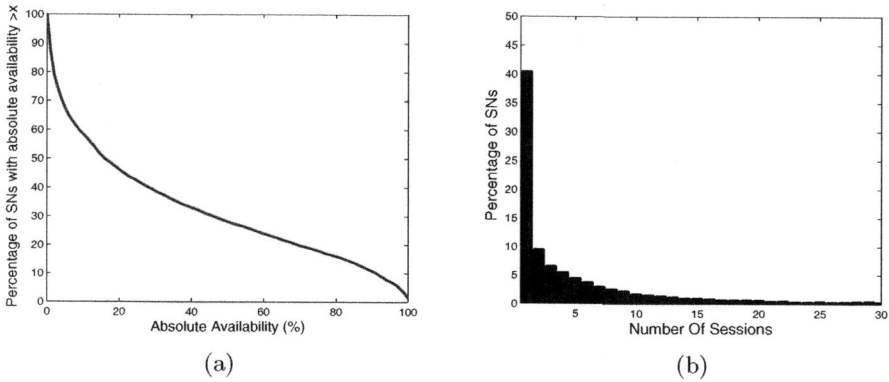

Fig. 2. (a) CDF of the absolute availability percentage of SNs. (b) Histogram of number of sessions in the residual life time of SNs.

4   Availability and Life Cycle of SNs

In this section we shed light on the life cycle of SNs. Naturally, a P2P system wishes to choose SNs that are available most of the time. We start by checking the **absolute availability** of SN, which is defined as *the time the SN is up during the test*. Figure 2(a) shows the absolute availability of SNs and surprisingly the percentage of absolute availability is low: 50% of the SNs are available less than 18% of the test time, i.e. less than 16 days. In order to understand this low absolute availability we start to analyze the life of SNs. Due to the test methodology we can not measure the life time of SNs but only the residual life time (recall that those SNs were already alive for unknown time when we start to measure their activity). Our first step is to understand the number of sessions an SN has during its life. Figure 2(b) shows the histogram of the number of sessions in the residual life of SNs. It is obvious that the number of SNs with only one session seems to be an exceptionally large number of 40%.

Motivated by this fact, and in order to further understand the behavior of SNs we define two groups of SNs: 1. The *Multiple Sessions in Life Time (MSLT)* SNs group - the group of SNs that have more than one session in their residual life time 2. Correspondingly, the *Single Session in Life Time (SSLT)* SNs group - the group of SNs that have exactly one session in their residual life time.[4]

Figure 3(a) shows the CDF of residual life time. One obvious outcome is that SNs that have only one session, have a very short residual life time. The median of residual life time at the SSLT group is 1.75 while the median of residual life time at the MSLT group is 67.5 days. SNs that were still alive when the experiment was ended (after 90 days) are clearly from the MSLT group: 38% of the MSLT SNs as opposed to 2.1% of the SSLT group. In the first 15 days the number of SNs that died permanently is the highest, and most of them are

---

[4] We note that our definition of an SN session refers only to sessions which were made by the SN while holding the same IP+port.

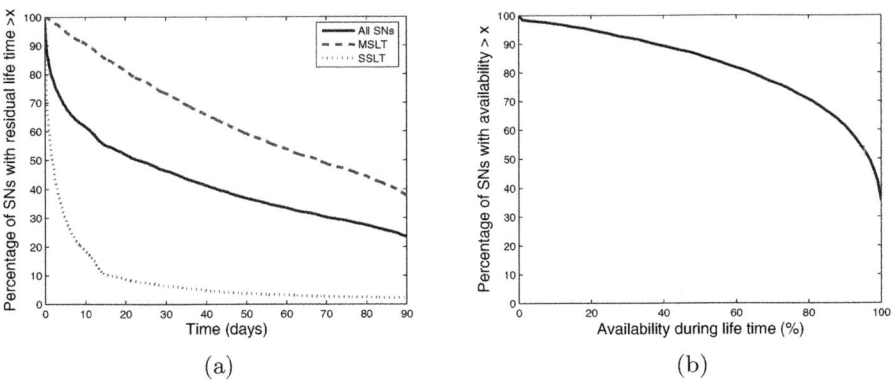

Fig. 3. (a) CDF of the residual life time. (b) CDF of the availability during life time of SNs.

from the SSLT group. From the graph shapes, it seems that the two groups have entirely different behavior, while the SSLT graph has exponential reduction in the first 15 days, the MSLT graph shows a steady linear reduction in all the test experiment period.

With this understanding we revisit the absolute availability definition and define new parameter the **availability during life** to be the *percentage of time a node is available during its residual life time.* Figure 3(b) shows that 90% of the nodes were available more than 37% of their life time. Note that, the SNs that were available 100% during their residual life, had only one session, i.e., they belongs to the SSLT group. Hence, we can now understand that the low measure of absolute availability of SN was mostly due to SNs that disappear permanently from the system, and not due to nodes that alternately leave and join the network.

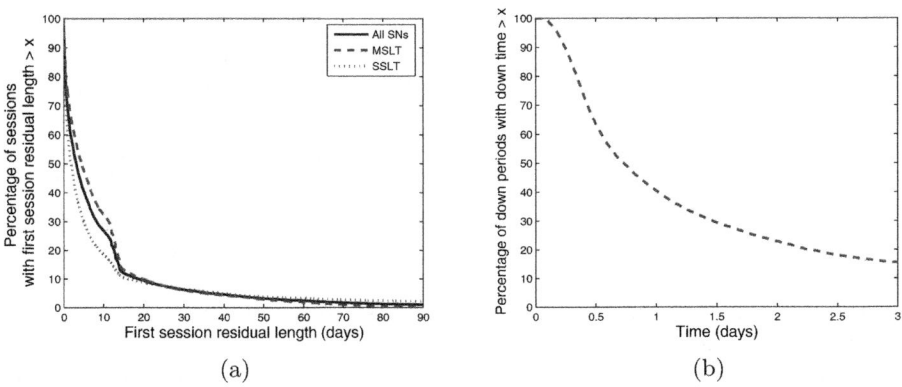

Fig. 4. (a) CDF of residual length of the first session. (b) CDF of downtime.

Motivated to understand the life cycle of the SNs we continue and analyze the length of the residual session length and down time between sessions (which is naturally applicable only for the MSLT group). One may predict that the length of the session will be similar between the two groups: i.e., that the MSLT SNs would have multiple sessions but the length of the session would be the same. Figure 4(a), analyzes the first session residual length, and shows that this is not the case: the SSLT has also shortest residual session length of 1.75 days in the median as opposed of 4.35 days median for the MSLT group. We explain this phenomenon in Section 6 after discussing the roots of those two groups. Figure 4(b) shows that the down time is relatively low to the session time; where the median of downtime is 0.8 days while the median of residual session length is 3 days (for all SNs) and 4.35 for the SNs in the MSLT group.

5   Churn and Accessibility of the P2P System

In this section, we discuss the impact of the life cycle of SNs on the system stability. We also show the impact on the stability if we would choose SNs only from SSLT group or only from MSLT group.

The accessibility of the P2P system - This is a new metric we suggest (somewhat similar to the group availability parameter of [8]). Motivation wise, this parameter correlates to the ability of a client that has an SN list which is T time old to access the P2P network using one of the live SNs on its list. Specifically, let us take a snapshot of the M active SNs of the P2P system at time t_0, the accessibility at time T of the set M is the percentage of the M SNs that are up also at time $t_0 + T$. Hence, if a client has a list of SNs with K SNs which were obtained T time ago, and the accessibility of the P2P system after T time is p, then the probability that a client can connect to the P2P system is $1 - (1 - p)^K$. Using this definition if an SN fails (sometime before time T) and then recovers at time T, the SN is useable as a node that never failed.

The accessibility parameter quantifies the ability of a client to access the P2P network with an old SN list. The lower accessibility value the higher the rate which the P2P system needs to update the SN list to ensure that the client can connect the network.

We assume that the P2P system cannot rely on bootstrapping servers (if they exist), that are in the premises of the P2P company, since they are fixed and known and hence vulnerable to blocking attempts of the P2P services. There are various reasons to block P2P systems. In the case of a P2P worm [26], such as STORM, the motivation of blocking the attempt is clearly to mitigate the spread of the worm. In order to design resilient worm, the accessability to the SNs that appear in the infected message should be high also after a long time (since in some cases a long time might elapse between the infection by the worm to the actual time the worm executes).

Legitime P2P services, are also in danger of being filtered, especially at enterprizes, that are concerned with data leak using encrypted P2P services that offer file transfers (such as Skype). Without being drawn into the legality aspects, an

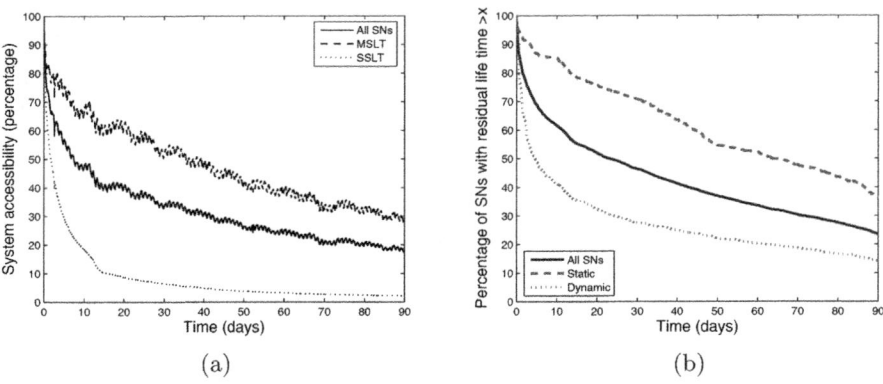

Fig. 5. (a) Accessibility of the SNs as function of time. (b) CDF of the residual life time of Static and Dynamic SNs.

ISP may wish to control the traffic of P2P services since the traffic consumes the ISP bandwidth. An ISP may also wish to filter or limit the rate of P2P services that compete with services offered by the ISP; for example, Skype may compete with ISP VoIP services.

In Figure 5(a) we show the percentage of accessibility to P2P network as a function of time elapsed from the beginning of our experiment. After 90 days of the test, the MSLT group has 38% accessibility, the SSLT group has accessability of 2.1% and the group of all the SNs has accessibility of 18%. I.e., choosing SNs only from the MSLT group would increase the acceptability by factor of 2.1. Note that there is no straightforward correlation between the accessibility of the system after T time and the life span and availability of the SN. Since if all the SNs had an availability of 50% and had a life span of 90 days it might still be that the accessibility of the system will be zero at fifty percentage of the time. In fact this can happen if all the SNs are correlated and in the same time zone: e.g, all of them are unavailable during the night. However, as Figure 5(a) shows this is not the case, and the graph shows only small waves, which we found were correlating to day and night zones in the USA. Overall the shape of the graph reassembles the graph of the residual life time (see Figure 2(b)). An analysis of the SNs origins according to countries, lead to the conclusion that this stability in accessibility is the direct outcome of the fact that the SNs are distributed over the whole world and over all time zones. We calculated the continents distribution of SNs that come from countries that contribute more than 1% to the total SNs and we have received that 38.24% are from North America, 29.92% from Europe and Africa, 16.02% from Asia and 15.82% that we did not classify. Hence we can conclude that the continent distribution of SNs, guarantees that there is a high chance that a Skype client will be able to connect to the network even if it was not alive for weeks.

The churn of a P2P system - Motivation wise the churn measures the number of times an SN goes down and we need to replace the SN. Specifically, we assume for simplicity that the system maintains a fixed number of SNs, M SNs. The assumption is that when an SN fails, the system picks another one to replace it. In this definition when a node fails and then recovers, from the system's perspective the node is like a new fresh node. We define churn as the number of SN turnovers (where turnover is when SN went down and another node was assigned to replace it) over a period of time T divided by M. For example, a churn of 2 per day means that on average the identity of each SN is replaced twice a day. Churn significantly influences the stability of a P2P system. An SN that goes down, requires the system to transfer all its functions to another SN. For example, in Skype, if the SN relays calls, all the calls need to be transferred to another SN.

We can estimate the Churn rate from the first session length. Recalling that churn is defined as the number of times we replaced an SN in order to maintain a fixed number of live SNs. Hence, when the session ends, we need to replace the SN with a new one. Let X be the random variable of the first session length. Then we can estimate the churn as $T/E(X)$, where T is a given time unit. However, note that we measured in our experiment the residual first session length (as presented in Figure 4(a)), and not the actual first session length, since we start to measure the nodes at some random time of their life. Let X_r be the random variable of the residual life of the first session, then $E(X_r) = E(X * Y)$ where Y is a random variable that indicates a percentage of the session time that has already elapsed when we start to measure the node. We use here a simplified assumption that Y is uniformly distributed and hence $E(X) = 2E(X_r)$[5]. Using this calculation we receive that the group of SSLT suffers from high churn of 0.35 turnovers per day compares to the group of MSLT with churn 0.22 turnovers per day. I.e., the SSLT churn is 1.66 higher than the MSLT churn. Moreover the churn of all the SNs group is 0.27, which means that choosing SNs only from the MSLT group will reduce the churn by 19%.

6   Dynamic Addresses and the Correlation to SSLT Group

Until now we have shown that there is a huge difference between the characteristics of SSLT and MSLT SNs (see summary of result at Table 1). We have also shown that if we choose the SNs only from the MSLT group it would improve the Churn and Accessibility of the P2P system dramatically. In this section we explain why these groups, SSLT and MSLT, are so different, and what the rational behind this partition into these two groups is.

We believe that the difference between SSLT and MSLT is inherent and it is related to the fact that some of the SNs belong to dynamic IP networks (usually

---

[5] Due to the paradox of residual life [27]- this calculation is not entirely accurate however using a more complicated and accurate calculation would not change significantly the result.

Table 1. Summary of the different characteristics of SSLT and MSLT SNs. The parameters are calculated using the 90 days of data. The residual life time, session length numbers are the median. The system accessability is calculated after the 90 days of the experiment.

	Residual life time	Residual first session length	System Churn	System Accessability
Total	22.7	3	0.27	18%
MSLT	67.7	4.35	0.22	38%
SSLT	1.75	1.75	0.35	2.1%

Table 2. Classification of the SNs according to their address type (Static/Dynamic)

	From All SNs	Static IP SNs	Dynamic IP SNs
Total	10,000	637	983
MSLT	59.7%	84.92%	38.55%
SSLT	40.3%	15.08%	61.45%
Ratio MSLT/SSLT	1.48	5.63	1/1.59

residential users connected by cable, xDSL and so on..). In this case an SN in the SSLT group is mostly SN with a dynamic IP and hence died since the IP address of the SN was replaced. However, there is a good chance that this SN is alive but with a different IP address[6]. This can explain also the exponential reduction in the residual life of SSLT SNs in the first days (see Figure 3(a)), since dynamic IPs live for only a few days [28]. As opposed to SSLT group, the MSLT group is composed from SNs with static IPs, that after leaving the network (for example closing the computer), can return with the same IP.

In order to support our assumption, we classified the IP to Static and Dynamic IPs using reverse DNS (rDNS, similar to the method of [28]). An rDNS maps an IP to its host name (e.g., "ip-66-186-253-215.dynamic.eatel.net"). We classify the IPs by searching in the returned results for keywords such as "static" and "dynamic". Using this method, we were able to classify 637 static IPs and 983 dynamic IPs. Note, that the rDNS technique was able only to identify only 15% of the SNs but the classification is almost 100% accurate. The SNs that were classified using the rDNS technique are a good random sample of the SNs. Moreover, rDNS is consider to be the most accurate technique. Other solutions, e.g. Spamhaus are able to classify all the IPs but are known to be accurate only 70% of the time [28]. Using Spamhaus we reach a similar result.

We present at Table 2 the correlation between the classification to SSLT and MSLT to Dynamic and Static IPs. It is clear that the vast majority, 84.92%, of the static IP SNs belongs to the MSLT group. With dynamic IPs group the result shows a weaker correlation with only 61.45% of dynamic IPs appeared in the SSLT group. Hence 38.55% of the dynamic IPs belong to the MSLT group.

---

[6] Dynamic IPs may occur also due to NAT or a cluster of proxies, however this is not relevant to Skype SN, since an SN can not be behind a firewall or NAT [9].

This is still a good indication that most of the dynamic IPs are from the SSLT group, if you take into account that the ratio between the number of SNs in the SNs group is 1.48 and for dynamic SNs group the ratio between MSLT to SSLT is 1/1.59.

We suspect that dynamic IPs that are in the MSLT group belong to a third group of IPs, to the "Sticky dynamic IP address" group. We believe that IPs that belong to this group are using DHCP. In the DHCP protocol [29], the ISP assigns its client an IP address and a lease time which determines the amount of time that subscriber can use this IP address.[7] In this period of time the client can also disconnect from the network and return to the same IP (hence the terminology "sticky dynamic IP address"). The lease time is usually a couple of days, since the ISP wishes to avoid load on the DHCP server and redundant traffic. We find support for our speculation in the fact that the Dynamic IPs that belong to the MSLT group had an average down time between sessions of 9 hours while static IPs that belong to MSLT had an average down time of 54 hours. We suspect that the dynamic IPs in the MSLT group had a relatively lower down time between sessions, since "Sticky dynamic IP address" can return to their original IP address only if the break is short and the lease time has not elapsed.

Another support for the correlation between the MSLT/SSLT groups to static/dynamic groups can be seen in the great similarity between the CDF graph of residual life time of the static/dynamic groups (see Figure 5(b)) to the CDF graph of residual life time of MSLT/SSLT groups (see Figure 3(a)).

The fact that MSLT/SSLT groups correlated to static/dynamic IP groups can explain the different residual life time and the different accessibility between the two groups. Dynamic IPs replace their IPs and hence live a shorter life time. However, one may wonder why there is also a difference in the first session length (see Figure 4(a)) which influences the Churn. Our speculation is that the root cause is the special type of clients that maintain static IPs. Static IPs are widely used for server applications or academic networks. Servers or academic networks need constant IPs since they need to be constantly up and available. A consequence of the required high availability is the relatively good infrastructure and hence the long session. We also observe, looking at the rDNS results, that there are a high number of university computers in the static group.

7 Conclusion and Discussion

Choosing which clients should become Super-Nodes in P2P networks is an important and crucial task for the stability of the network. As far as we know we are the first paper that shows the impact of choosing static IPs on the different aspects of network stability. The high stability of static IP is due to two reasons: the fact that the static IP does not change the IP address, and the fact that computer which is connected through a static IP connection is relatively more stable, since the computer is usually used as a server. The fact that static IPs

---

[7] During the lease time the subscriber can also renew the address lease time.

are more stable can be used to choose stable SNs, and thus increase the stability of the P2P network. While it is a hard task to classify the type of address (static or dynamic) using the IP address alone, this is an easy task for the P2P application.

Specifically, the P2P application is usually designed in such a way that the application in the clients send from time to time keepalive messages to the central unit of the P2P network with some identifer of the client application. Using those keepalive messages the P2P network can detect the changes in the IP addresses of the computer. Change of IP can be due to the roaming of the computer (incase of laptop or smart phone device) or due to the fact that the IP address of the computer is a dynamic IP address. A thumb rule suggests that if the change is to another IP in the same subnetwork (usually the /24 or according to the BGP prefixes [30]) and the change is periodically then the change is due to the fact that this is dynamic IP [28]. Hence our paper observation on the rule of type of address (dynamic/static) is an operative guideline to the designers of P2P system.

Our measurements on the life cycle of SNs reveal that there is a set of nodes which is very stable; that its session duration is a couple of days and its life span is over 3 months (note that 38% of the nodes in the MSLT group live during the entire experiment which was 3 months). Those nodes can be used as an important building block in designing a more stable P2P network.

Acknowledgment

The authors would like to thank Amir Lev, Commtouch's CTO, Prof. Hanoch Levy from Tel-Aviv University and Omer Dekel, a graduate student at IDC, for their helpful suggestions.

References

1. Garcés-Erice, L., Biersack, E.W., Felber, P., Ross, K.W., Urvoy-Keller, G.: Hierarchical peer-to-peer systems. In: Kosch, H., Böszörményi, L., Hellwagner, H. (eds.) Euro-Par 2003. LNCS, vol. 2790, pp. 1230–1239. Springer, Heidelberg (2003)
2. Wang, F., Liu, J., Xiong, Y.: Stable peers: Existence, importance, and application in peer-to-peer live video streaming. In: IEEE Infocom (2008)
3. Saroiu, S., Gummadi, S., Gribble, P.: A measurement study of peer-to-peer file sharing systems. In: Multimedia Computing and Networking (2002)
4. Sen, S., Wang, J.: Analyzing peer-to-peer traffic across large networks. In: ACM SIGCOMM Internet Measurement Workshop (2002)
5. Bhagwan, R., Savage, S., Voelker, G.: Understanding availability. In: Kaashoek, M.F., Stoica, I. (eds.) IPTPS 2003. LNCS, vol. 2735, Springer, Heidelberg (2003)
6. Gummadi, K.P., Dunn, R.J., Saroiu, S., Gribble, S.D., Levy, H.M., Zahorjan, J.: Measurement, modeling, and analysis of a peer-to-peer file-sharing workload. In: ACM SOSP (2003)
7. Godfrey, P., Shenker, S., Stoica, I.: Minimizing churn in distributed systems. In: ACM SIGCOMM (2006)

8. Stutzbach, D., Rejaie, R.: Characterizing churn in peer-to-peer networks. Tech. Rep. (May 2005)
9. Guha, S., Daswani, N., Jain, R.: An experimental study of the skype peer-to-peer voip system. In: IPTPS (2006)
10. Baset, S.A., Schulzrinne, H.G.: An analysis of the skype peer-to-peer internet telephony protocol. In: IEEE INFOCOM (2006)
11. kazaa (2007), http://kazaa.com
12. Joost, http://joost.com
13. iMesh, http://www.imesh.com/
14. Morpheus discussion group, http://www.gnutellaforums.com/morpheus-windows/
15. Suh, K.F., Kurose, D.R., Towsley, D.J.: Characterizing and detecting skype-relayed traffic. In: IEEE INFOCOM (2006)
16. Ehlert, S., Magedanz, T., Petgang, S., Sisalem, D.: Analysis and signature skype voip session traffic. Tech. Rep. NGNI-SKYPE-06B (2006)
17. Fabrice, D.: Skype uncovered (2005), http://www.ossir.org/windows/supports/-2005/2005-11-07/EADS-CCR_Fabrice_Skype.pdf
18. Recon 2006 - Fabrice Desclaux and Kostya Kortchinsky - Vanilla Skype (2006), http://recon.cx/en/f/vskype-part1.pdf, http://recon.cx/en/f/vskype-part2.pdf
19. Biondi, P., Desclaux, F.: Silver needle in the skype. In: Blackhat (2006)
20. Cicco, L.D., Mascolo, S., Palmisano, V.: An experimental investigation of the congestion control used by skype voIP. In: Boavida, F., Monteiro, E., Mascolo, S., Koucheryavy, Y. (eds.) WWIC 2007. LNCS, vol. 4517, pp. 153–164. Springer, Heidelberg (2007)
21. Apoc Matrix, The SkypeLogger (December 2006), http://www.epokh.org/drupy/files/Skype%20Logger.pdf
22. Chen, K.-T., Huang, C.-Y., Huang, P., Lei, C.-L.: Quantifying skype user satisfaction. In: ACM SIGCOMM (2006)
23. Hofeld, T., Binzenhöfer, A., Fiedler, M., Tutschku, K.: Measurement and analysis of skype voip traffic in 3g umts systems. In: The 4th International Workshop on Internet Performance, Simulation, Monitoring and Measurement (2006)
24. Wang, X., Chen, X., Jajodia, S.: Tracking anonymous peer-to-peer voip calls on the internet. In: ACM Conference on Computer and Communications Security (2005)
25. Bremler-Barr, A., Dekel, O., Levy, H.: Harvesting skypesuper-nodes. Tech. Rep. (May 2008)
26. Holz, T., Steiner, M., Dahl, F., Biersack, E., Freiling, F.: Measurements and mitigation of peer to-peer-based botnets: A case study on storm worm. In: LEET 2008: The First USENIX Workshop on Large-Scale Exploits and Emergent Threats (2008)
27. Kleinrock, L.: Queueing Systems: Volume I – Theory. Wiley Interscience, New York (1975)
28. Xie, Y., Yu, F., Achan, K., Gillum, E., Goldszmidt, M., Wobber, T.: How dynamic are ip addresses. In: ACM SIGCOMM (2007)
29. Khadilkar, M., Feamster, N., Sanders, M., Clark, R.: Usage-based dhcp lease time optimization. In: Internet Measurement Conference (2007)
30. Krishnamurthy, B.: On network-aware clustering of web clients. In: Proceedings of ACM SIGCOMM, pp. 97–110 (2000)

Reduce to the Max: A Simple Approach for Massive-Scale Privacy-Preserving Collaborative Network Measurements (Short Paper)

Fabio Ricciato[1] and Martin Burkhart[2]

[1] University of Salento, Italy and FTW, Austria
[2] ETH Zurich, Switzerland

Abstract. Privacy-preserving techniques for distributed computation have been proposed recently as a promising framework in collaborative inter-domain network monitoring. Several different approaches exist to solve such class of problems, e.g., Homomorphic Encryption (HE) and Secure Multiparty Computation (SMC) based on Shamir's Secret Sharing algorithm (SSS). Such techniques are complete from a computation-theoretic perspective: given a set of private inputs, it is possible to perform arbitrary computation tasks without revealing any of the intermediate results. In this paper we advocate the use of "elementary" (as opposite to "complete") Secure Multiparty Computation (E-SMC) procedures for traffic monitoring. E-SMC supports only simple computations with *private input and public output*, i.e., they can not handle secret input nor secret (intermediate) output. The proposed simplification brings a dramatic reduction in complexity and enables massive-scale implementation with acceptable delay and overhead. Notwithstanding their simplicity, we claim that a simple additive E-SMC scheme is sufficient to perform many computation tasks of practical relevance to collaborative network monitoring, such as anonymous publishing and set operations.

1 Introduction

Privacy-preserving techniques for distributed computation have been proposed recently to support inter-domain collaborative network monitoring [1]. In the reference scenario, a set of collaborating ISPs are unwilling to share local traffic data due to business sensitivity, but they have a collective interest to perform some operation on such data (e.g., aggregation) and share the final result. For example, they might want to aggregate local traffic measurements in order to reconstruct global statistics, which are further processed in order to unveil global threats (e.g., botnets) or discover macroscopic anomalies. As pointed out in [2], each ISP would benefit from comparing its local traffic conditions with the global view aggregated over all other ISPs, especially in the occasion of anomalies and alarms, in order to hint at whether the (unknown) root cause is local or global. Also, ISPs might be ready to share with other ISPs information about security incidents observed locally (e.g., intrusion alarms) provided they can do so anonymously.

Two possible approaches to solve such class of problems are Homomorphic Encryption (HE) and Secure Multiparty Computation (SMC) based on Shamir's Secret Sharing

J. Domingo-Pascual, Y. Shavitt, and S. Uhlig (Eds.): TMA 2011, LNCS 6613, pp. 100–107, 2011.
© Springer-Verlag Berlin Heidelberg 2011

algorithm (SSS for short). Both such techniques are "complete" from a computation-theoretic perspective (see [3] and [4]): given a set of *private inputs*, it is possible, in principle, to compute any arbitrary function, including structured algorithms involving conditional statements, without revealing any of the intermediate results. In fact, a distinguishing feature of HE and SSS is that they can operate also on *secret inputs* and/or provide *secret outputs* (see Fig. 1(a)). The notions of *secret* and *private* are distinct: *private data* is known in cleartext to at least one player (and usually only to one), while *secret data* remains unknown by all players and can not be reconstructed unless a minimum number of players agree to do so. On the other hand, such techniques are computationally expensive — especially HE — and therefore do not scale well in the number of players and/or in the rate of computation tasks (queries).

Here we advocate the use of "elementary" (as opposite to "complete") SMC procedures for collaborative traffic monitoring. Such techniques — hereafter referred to as E-SMC for short — support only simple computations with *private inputs and public output*, i.e., they can not handle secret input nor secret (intermediate) output. We show that such a simplification allows for an enormous reduction in computational complexity and overhead, making such techniques amenable to massive-scale implementation. Notwithstanding their simplicity, we claim that E-SMC is sufficient to perform a broad variety of tasks of practical importance in the field of collaborative traffic monitoring. In fact, queries can be chained to build more structured computation tasks (ref. Fig. 1(b)) whenever intermediate results, which are always public in E-SMC, are not regarded as sensitive. Moreover, we show that an additive E-SMC scheme can be combined with local transformations on the private data and/or with particular data structures (e.g., Bloom Filters, bitmap strings) in order to extend the range of supported operations. Thanks to their simplicity, collaborative systems based on E-SMC are amenable to massive-scale implementation, with very large number of players and/or very high rate of queries.

In this initial work we take a first step towards unfolding the potential of E-SMC for traffic monitoring. We make three main contributions. First, we present a simple scheme for E-SMC, called GCR, which is based on additive-only or multiplicative-only secret computation and extends an idea presented earlier in [5]. Second, we highlight some system-design aspects of GCR that enable massive-scale implementation: in particular, we propose to split the computation into *offline* randomization and *online* aggregation phases. Third, we present a number of use-cases and operations relevant to collaborative traffic monitoring and show how they can be mapped to E-SMC queries in combination with Bloom Filters and bitmap strings.

2   The GCR Method

We consider the classical SMC scenario where a set of n players collaborate to compute a function of some private data — e.g., traffic statistics, network logs, records of security incidents. As customary in SMC, we assume a *semi-honest* model (also known as *honest-but-curious*): all players cooperate honestly to compute the final result, but a subset of them might collude to infer private information of other players. In other words, no *malicious* player will attempt to interrupt nor corrupt the computation process, e.g., by providing incorrect or incomplete input data.

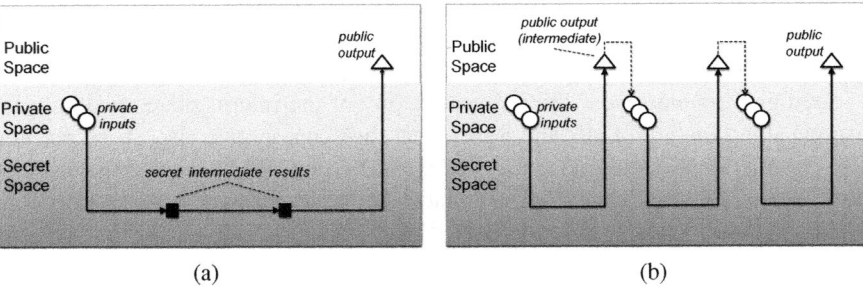

Fig. 1. Graphical representation of a "complete" secure procedure with secret intermediate results (a) and a sequence of "elementary" secure operations chained by public intermediate results (b)

Hereafter we present a simple method to perform secure private *addition* which extends an idea presented earlier by Atallah et al. in [5, §4.1] based on additive secret sharing. We refer to this method as "Globally-Constrained Randomization" (GCR for short). We show also that GCR, which is simple conceptually, lends itself to massive-scale implementation. A variation of the scheme to perform secure *multiplication* is presented in the extended version of this work [6], where the interested reader can find also a detailed comparison between SSS and GCR.

Notation. We consider a set of $n \geq 3$ players $\{P_i, i = 1 \ldots n\}$. The maximum number of colluding players is denoted by l (collusion threshold) with $l \leq n - 2$. For each computation task (query) each player P_i involves the following elements:

- a_i is the *private input* of P_i to the summation. For some queries, it is obtained by applying a local transformation $g()$ on some other inner private data b_i, i.e., $a_i = g(b_i)$.
- r_i is the private *random element* which P_i has previously generated cooperatively with other players in the way presented later.
- $v_i \stackrel{\text{def}}{=} a_i + r_i$ is the *public input* which P_i eventually announces to the other players.

The collection of random elements across all players constitutes a Random Set (RS) and will be denoted by $\mathbf{r} \stackrel{\text{def}}{=} \{r_i, i = 1 \ldots n\}$. The goal of the computation round is to obtain the *public output* result $A \stackrel{\text{def}}{=} f(a_1, a_2 \ldots a_n) = f(g(b_1), g(b_2), ..g(b_n))$ without disclosing the values of the individual a_i's. For each computation, all input elements (a_i, r_i, v_i) and the output A must be in the same format, defined over the same *additive commutative group* (Abelian group). We will consider the following distinct cases:

Scalars: a_i, r_i and A are real or integer numbers defined in the interval $\mathbb{R}_p \stackrel{\text{def}}{=} [0, p]$. For the sake of simplicity we will assume p integer, but not necessarily prime. The group operation in this case is modulo-p addition. A generic random element x is a random value extracted uniformly in $[0, p]$, i.e., $x \sim \mathcal{U}(0, p)$. The null element is the zero value. For integers, it is convenient to choose $p = 2^q$ with integer q so that modulo-p addition maps to wrap-around of a q-bit counter.

Binary strings: a_i, r_i and A are binary strings of length k. The group operation is therefore bitwise addition (XORing). In this context a generic random element x is a random string, i.e., a collection of bits set randomly to 1 or 0 independently and with equal probabilities. The null element is a string with all '0's.

Arrays of counters: a_i, r_i and A are vectors of k elements, each element being a $q-$bit counter. The group operation is an array of k parallel modulo$-p$ additions. A generic random element is a collection of k random values $\langle x_1, x_2, ..x_k \rangle$ extracted independently and uniformly in $[0, p-1]$. The null element is an array of zeros.

The format of input elements a_i, r_i and, if applicable, the choice of the transformation function $g()$ depend on the type of operation (query) as detailed in §3. We adopt the symbols '+' and '$\sum$' to refer generically to the addition between two or multiple terms, without specifying the group operation.

Overview. The central aspect of GCR is that the RS is constructed in a way that guarantees the zero-sum condition, i.e., the composition of random elements across *all* users sums up to the null element, formally $\sum_{i=1}^{n} r_i = \mathbf{0}$. Moreover, the generation of RS ensures that the individual r_i's can not be inferred by other players, provided that the number of colluding players remains below the threshold l. Each player P_i then shares with the others (e.g., via a central collector) the sum of data plus random elements, i.e., $v_i = a_i + r_i$, which serves as the public input to the computation. When *all* input elements v_i are collected, the value of A is obtained simply by the total sum, formally: $\sum_{i=1}^{n} v_i = \sum_{i=1}^{n} (a_i + r_i) = \sum_{i=1}^{n} a_i + \sum_{i=1}^{n} r_i = A + 0 = A$. Note that the value of A can be reconstructed only when the inputs from *all* players have been collected: it is sufficient that a single player (among those that have contributed to generate the RS **r**) fails to provide its input element to block the computation of A. This is the main disadvantage of GCR compared to SSS, as discussed more in details in [6].

Generation of Random Sets. Hereafter we describe how each generic player P_i ($i = 1 \ldots n$) constructs its random element r_i in cooperation with other players, so as to collectively build the RS **r**. Note that the RS generation procedure is completely asynchronous and can be run *in parallel* by all players. Each random element is initially set to the null element, i.e., $r_i = 0$. Each player P_i extracts $l + 1$ random variables $x_{i,j}$ ($j = 1 \ldots l + 1$) and computes their sum $y_i \stackrel{\text{def}}{=} \sum_j x_{i,j}$. It calculates the additive inverse[1] $\overline{y_i}$ of y_i and adds it to its own random element, i.e., $r_i \leftarrow r_i + \overline{y_i}$. At the same time, P_i contacts $l + 1$ randomly selected other players and sends one variable $x_{i,j}$ to each of them: each contacted player P_j will then increment its random element by $x_{i,j}$, i.e., $r_j \leftarrow r_j + x_{i,j}$. This method is secure against collusion of up to l players. Note that l is a free parameter, independent from system size n, which can be tuned to trade-off communication overhead with robustness to collusion — both scale linearly in l.

Computation phase. With GCR the computation is basically a summation over n public inputs, the v_i's, and no particular constraint applies to the aggregation method, which can be centralized or distributed. In other words, the GCR method is oblivious to the adopted input aggregation scheme. For the sake of simplicity, we assume here a fully centralized scheme, with a single master in charge of launching the query, collecting the n public inputs, computing the result and finally publishing it to all the players.

---

[1] In modular arithmetic the additive inverse $\overline{y}$ of y is the element that satisfies $\overline{y} + y = 0$. For real numbers in $[0, p], \overline{y} = p - y + 1$, while for binary strings $\overline{y} = y$.

Decoupling RS generation and computation. One key advantage of GCR is that the process of generating the RS is completely decoupled — and can be run independently — from the actual computation round. We devise a system where lists of RS are generated *offline* and stored for later use. At any time, each player P_i has available a collection of random elements $r_i[u]$, indexed in u, which can be readily used for future computation rounds. The communication protocol must ensure that the RS indexing is univocal and synchronized across all players. During the *online* computation phase, the query command broadcasted by the central master will indicate explicitly the RS index to be used for the production of the public inputs v_i's.

Performing RS generation offline brings several advantages, especially in massive-scale systems. First, it minimizes the query response delay down to the same value of an equivalent cleartext summation. Second, generation of multiple RS can be *batched*, meaning that in a single secure connection (typically SSL over TCP) two players i, j can exchange multiple $\langle$variable,index$\rangle$ pairs $\{x_{i,j}[u], u\}_u$ which collectively build a collection of RS $\{\mathbf{r}[u]\}_u$. This reduces dramatically the communication overhead for connection establishment (handshaking, authentication, etc.). Moreover, the RS generation process can be scheduled in periods of low network load, e.g., at night.

In GCR the set of players participating in the computation round must match *exactly* the set of players that have previously built the RS. If RSs are generated offline, the set of players might have changed between the generation of $\mathbf{r}[u]$ and its consumption in a query. It would be very impractical to trash all pre-computed RSs upon every new player joining or leaving the system — an event not infrequent if the number of players is large. In order to ensure consistency, the legacy RS must be incrementally adjusted upon join or leave of players, but that requires at most $l + 1$ operations performed by the joining/leaving player (see [6] for additional details).

3 Advanced Operations

The GCR scheme can be used directly to perform basic additive tasks, such as aggregation and counting (refer to [6, §4]). Here we show a few examples of more advanced operations which can be mapped to additive E-SMC queries — and as such can be supported by GCR — in combination with specific constraints on the input data elements and/or a proper local transformation function $g()$. For each of them we illustrate a possible application for collaborative network monitoring. This section is one of the main contributions of the paper: to the best of our knowledge we are the first to "interpret" the following operations as applications of SMC using the additive sharing scheme.

3.1 Set Operations

In this section, we first describe how (probabilistic) set operations can be implemented using bloom filters with any SMC scheme that supports both, private additions and multiplications (e.g., SSS). We then outline what subpart of that functionality can easily be implemented with GCR.

Bloom filters (BF) are powerful data structures for representing sets [7]. A BF representing a set $S = \{x_1, x_2, \ldots, x_n\}$ of n elements is described by an array of m bits, initially all set to 0. The BF uses k independent hash functions $h_1, \ldots, h_k$ with range $1, \ldots, m$. For each element $x \in S$, the bits $h_i(x)$ are set to 1 for $1 \leq i \leq k$. For checking whether an element y is a member of S, we simply check whether all bits $h_i(y)$ are set to 1. As long as the BF is not saturated, i.e., m is chosen sufficiently large to represent all elements, the total number of non-zero buckets allows to accurately estimate $|S|$. Counting Bloom Filters (CBF) are a generalization of BFs, which use integer arrays instead of bit arrays. Thus, CBFs allow to represent *multisets*, in which each element can be represented more than once. Note that a (C)BF allows to efficiently check for element membership, but not to enumerate the contained elements. Compared to state-of-the-art approaches for privacy-preserving set operations via HE (e.g., [8]), the use of (C)BF allows for very efficient and scalable solutions.

Set Union. If each player i has a local set S_i, they can construct the union of their sets $S = S_1 \cup S_2 \cup, \ldots, \cup S_n$ by performing private OR ($\vee$) over their BF arrays. If inputs are multisets, represented by CBFs, the aggregation operation is addition instead of OR. Using CBFs, each player can learn the number of occurrences of specific elements across all players or the number of other players that report each element (by using a BF as input). From the aggregate CBF, one could, for instance, compute the entropy of the empirical element distribution.

Set Intersection. In order to perform set intersection on BFs, the players simply use the AND ($\wedge$) operation for aggregating their sets $S = S_1 \cap S_2 \cap, \ldots \cap S_n$. Only buckets set to 1 in all the players' BFs will evaluate to 1 in the aggregate BF. In this specific scenario, it is also possible for each player i to enumerate all elements in S simply by iterating over all $x \in S_i$ and checking whether $x \in S$, since $S \subseteq S_i$.

Set Operations with GCR. GCR directly supports the addition operation and therefore set union on multisets. If the counts in each bucket are not sensitive, the union and intersection of sets can be computed from the public union of multisets — the intersection, for instance, is given by selecting all elements with count n. However, private union and intersection directly on sets can not be delivered by GCR. In fact, union requires OR, i.e., a combination of addition and multiplication not supported by GCR, while the problem with intersection is that multiplicative GCR does not include 0 (see [6]).

3.2 Anonymous Publishing

The goal is to let one player P_1 publish to all other players a binary string w without revealing its identity. The string w can represent, for example, a piece of malware payload that P_1 has discovered with an IDS, or the description of a security incident which was observed locally. Moreover, w could be used as a public condition for a future counting round, e.g., to discover how many other players have observed the same event. There are several reasons why the publisher wants to remain anonymous. First, knowing that it was hit by the malware might be detrimental to its reputation among customers. Second, such information might benefit other potential attackers.

DC-nets [9] are a basic and unconditionally secure solution for anonymous publishing. In the following, we devise an alternative solution that does not require pair-wise shared secrets, and deals with the problem of detecting and/or avoiding collisions.

Let k denote the length of string w, and denote by $C(w)$ a Cyclic Redundancy Check (CRC) control field of length c computed on w — the need for CRC is explained below. It is straightforward to map an Anonymous Publishing round to a bit-wise summation on strings of length $k + c$. The publisher P_1 sets its data element to the concatenation of w and $C(w)$, i.e., $a_1 = \langle w, C(w) \rangle$, while all other players set their data elements to null ($a_j = 0$, $j \neq 1$). Therefore the public result will return the string w in cleartext, i.e., $A = a_1 = \langle w, C(w) \rangle$, but since the individual data elements remain unknown the identity of the publisher is protected. Such a simple approach works only if exactly *one* player publishes in the computation round: if two (or more) players P_1 and P_2 attempt to publish different strings, there is a collision — i.e., the computed result will be the combination $A = \langle w_1 \oplus w_2, C(w_1) \oplus C(w_2) \rangle$ ('$\oplus$' for bit-wise summation) from which neither of the elements w_1, w_2 can be derived. However the collision can be easily revealed by CRC failure as in general $C(w_1 + w_2) \neq C(w_1) \oplus C(w_2)$. The "collision recovery" procedure can simply foresee the repetition of new anonymous publishing rounds associated to a back-off scheme to avoid that the same players collide again in the next round — a mechanism conceptually equivalent to Slotted-Aloha.

A simple "detection and recovery" approach is not effective when the instantaneous rate of publishing attempts is high — this is of particular concern in large-scale system with many players ($n >> 1$) and/or in presence of correlated attempts (e.g., a spreading malware payload caught simultaneously by different domains). In such cases it is preferable to adopt a "collision prevention" method by orderly scheduling the publishing rounds for different players. This can be achieved by a single round of anonymous scheduling, as explained below.

3.3 Anonymous Scheduling

The problem is defined as follows: out of the total n players, a subset of $m < n$ "active" players are ready to perform a given action, e.g., anonymous publishing. The goal is to schedule the m active players *without knowing nor revealing their identities*. This apparently difficult task can be easily accomplished by bit-wise summation over strings of size $k >> m$. At the query round, the inactive players set their data elements to the null string, while each *active* player P_i extracts uniformly a random integer $q_i \sim \mathcal{U}(1, k)$ and then builds its data element a_i with a single '1' at the q_i-th position and all other bits set to '0'. The bitmap length k must be large enough to ensure that bit-collision probability — i.e., two or more players independently picking the same random value q_i — is kept acceptably low. Assuming that no bit-collision has occurred, the final (public) result A is a bitmap with m '1's and $k - m$ '0's. Upon learning A, each active player P_i checks whether the bit in the q_i position is set to '1', and if so it counts the number of '1's in the preceding positions, say μ_i, from which he learns it has been scheduled in the successive $(\mu_i + 1)$-th query round. If otherwise the q_i-th bit is '0', P_i infers that a collision has occurred and waits for the next scheduling round.

In case of bit-collisions the round does not completely fail: if collisions involves only two (or any even number of) players, the colliding players will simply wait for the next scheduling query. If three (or any odd number of) players have collided on the same q-th bit, they would again collide in the q-th query round. However this is not a serious problem as far as collisions in the query rounds can be detected and recovered (e.g., by CRC failure in case of Anonymous Publishing). An alternative strategy is to preliminarily discover the exact number of active players m via a simple counting query, and then validate the scheduling round only if the number of '1' equals m (see [6, §5.4]).

4   Conclusions

In this initial work we have introduced the distinction between "complete" and "elementary" SMC techniques. It was shown that an elementary additive scheme, namely GCR, lends itself to massive-scale implementation through the separation between randomization and computation phases. Despite its simplicity, GCR is sufficient to support several common operations for collaborative inter-domain monitoring. Furthermore, it can be combined with Bloom Filters and/or bitmap strings to handle set operations, anonymous scheduling and anonymous publishing (e.g., of security alarms). GCR allows to leverage SMC in collaborative inter-domain monitoring systems with very high rate of queries and/or large number of players.

Acknowledgment. This work was supported by the DEMONS project funded by the EU 7th Framework Programme [G.A. no. 257315] (http://fp7-demons.org).

References

1. Roughan, M., Zhang, Y.: Secure distributed data-mining and its application to large-scale network measurements. ACM Computer Communication Review 36(1) (2006)
2. Burkhart, M., Strasser, M., Many, D., Dimitropoulos, X.: SEPIA: Privacy-Preserving Aggregation of Multi-Domain Network Events and Statistics. In: 19th USENIX Security Symposium, Washington, DC, USA (August 2010)
3. Gentry, C.: Fully homomorphic encryption using ideal lattices. In: ACM symposium on Theory of Computing. ACM, New York (2009)
4. Ben-Or, M., Goldwasser, S., Wigderson, A.: Completeness theorems for non-cryptographic fault-tolerant distributed computation. In: ACM Symp. on Theory of computing, STOC (1988)
5. Atallah, M., Bykova, M., Li, J., Frikken, K., Topkara, M.: Private collaborative forecasting and benchmarking. In: Proc. ACM WPES 2004 (October 2004)
6. Ricciato, F., Burkhart, M.: Reduce to the max: A simple approach for massive-scale privacy-preserving collaborative network measurements (extended version). CoRR, vol. abs/1101.5509 (2011), http://arxiv.org/abs/1101.5509
7. Broder, A., Mitzenmacher, M.: Network applications of bloom filters: A survey. Internet Mathematics 1(4), 485–509 (2004)
8. Kissner, L., Song, D.: Privacy-Preserving Set Operations. In: Shoup, V. (ed.) CRYPTO 2005. LNCS, vol. 3621, pp. 241–257. Springer, Heidelberg (2005)
9. Chaum, D.: The dining cryptographers problem: Unconditional sender and recipient untraceability. Journal of Cryptology 1(1), 65–75 (1988)

An Analysis of Anonymity Technology Usage

Bingdong Li[1,*], Esra Erdin[1,*], Mehmet Hadi Güneş[1],
George Bebis[1], and Todd Shipley[2]

[1] University of Nevada - Reno, Reno, NV 89557
{bingdonli,eerdin,mgunes,bebis}@cse.unr.edu
[2] Vere Software, Reno, NV 89519
todd@veresoftware.com

Abstract. Anonymity techniques provide legitimate usage such as privacy and freedom of speech, but are also used by cyber criminals to hide themselves. In this paper, we provide usage and geo-location analysis of major anonymization systems, i.e., anonymous proxy servers, remailers, JAP, I2P and Tor. Among these systems, remailers and JAP seem to have minimal usage. We then provide a detailed analysis of Tor system by analyzing traffic through two relays. Our results indicate certain countries utilize Tor network more than others. We also analyze anonymity systems from service perspective by inspecting sources of spam e-mail and peer-to-peer clients in recent data sets. We found that proxy servers are used more than other anonymity techniques in both. We believe this is due to proxies providing basic anonymity with minimal delay compared to other systems that incur higher delays.

Keywords: Anonymizer, onion routing, Tor.

1 Introduction

Anonymizers are services that enable users of the Internet to browse the web anonymously. They allow a user to maintain a level of privacy that prevents the collection of identifying information such as the IP address while surfing on the web. Anonymizers are an offspring of mix networks that use a chain of proxy servers to create hard-to-trace communications [4]. These anonymity services are provided by either commercial companies driven by subscription fees, noncommercial organizations profiting from advertising, or home-brewed services through open source anonymous tools. Community contributed systems include The Onion Router (Tor) [6], the Invisible Internet Project (I2P) [1], and the Java Anon Proxy (JAP) [2].

Anonymity is defined as a state in which an agent is not identifiable within an anonymity set [12, 15, 17]. The anonymity set is a system of senders, receivers, and servers in the communication network. Anonymity is a combination of both *unidentifiability*, i.e., observers can not identify any individual agent, and *unlinkability*, i.e., observers can not link an agent to a specific message or action.

---

* Equally contributing authors.

J. Domingo-Pascual, Y. Shavitt, and S. Uhlig (Eds.): TMA 2011, LNCS 6613, pp. 108–121, 2011.
© Springer-Verlag Berlin Heidelberg 2011

Anonymity has always been a dichotomous issue in both social life and cyber space. Anonymity technologies have been used for criminal purposes as well as legitimate purpose. On one side, anonymous technologies provide legitimate usages such as privacy and freedom of speech, anti-censorship, anonymous tips for law enforcement, and surveys such as evaluation and feedback. On the other side, anonymous technologies provide protection to criminals in facilitating online crimes such as spam, piracy, information and identity theft, cyber-stalking and even organizing terrorism. Additionally, they may be utilized for Internet abuse for bypassing the Internet use policy of an organization, exposing organization to malicious activities, abusing organization resources, and prevent web filters from monitoring.

Anonymizer systems send data packets over randomly chosen relays so that no single system has information about both the sender and the receiver. Since many users use these intermediaries at the same time, the Internet connection of any one single user is hidden among the connections of all other users. Hence, no individual system, internal or external, can determine which connection belongs to which user. Anonymity research remains a very active area where investigators have focused on anonymous communication, traffic analysis, provable shuffles, anonymous publications, private information retrieval, formal methods, communication censorship, and traffics [5, 7, 12].

In this paper, we analyzed usage of popular anonymity systems including anonymity proxy servers, remailers, JAP mix network, I2P and Tor. For this study, we collected the server lists of each technology and looked up the geolocation of servers. During our exploration, we identified 1,441 anonymity proxy servers, 15 remailers, 11 JAP mixers, 483 I2P relays, and 10,510 Tor relays. We observed that U.S. and Germany were among the top 5 server providers for proxy, Tor and I2P systems and additionally France and Russia were among the top 5 for Tor and I2P systems.

We then performed a detailed analysis of Tor system, the most popular anonymity system, by setting up two servers to analyze Tor usage. During the experiment our servers relayed 150GB of traffic. In this experiment, we observed that relays from Germany and U.S. contribute most bandwidth resources to Tor system and that they have the highest number of Tor users.

Finally, we analyzed anonymity systems from service perspective by analyzing spam e-mail and peer-to-peer client sources of recent data sets. In spam data, we observed e-mails sent through commercial anonymizer services such as GoTrusted. Moreover, we found that proxy servers are used more than other anonymity techniques by spammers and peer-to-peer users to hide their IP addresses. We believe this is due to proxies providing basic anonymity with minimal delay compared to other systems that incur higher delays.

In the rest of the paper, we first analyze well known deployed anonymity systems in Section 2. In Section 3, we analyze the usage of Tor anonymity system in depth. In Section 4, we analyze anonymity system usage in different networks. Related work is discussed in Section 5. Finally, we provide our conclusion in Section 6.

2   Analysis of Anonymization Techniques

There are many categories of anonymity systems. From a usability point of view, anonymous communication can be classified in two categories: *high latency systems*, mostly used by email anonymity that provide strong anonymity, and *low latency systems*, mostly used by anonymous web browsing that have better performance. Other categories can be based on trust level, network type, anonymity properties, or adversary capability.

In this section, we review well-known deployed anonymity systems and provide geographic distribution of their servers.

2.1   Proxy Server

A proxy server is the easiest to deploy anonymity system mostly used for low latency browser anonymity [7]. The basic idea behind a proxy server is that a client uses a proxy server to surf the web as in Figure 1. The proxy server performs client requests using the proxy server's identity rather than the client's real iden-

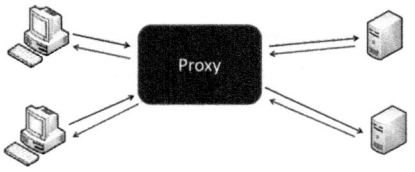

Fig. 1. Proxy Server

tity. Proxy servers relay requests from users to their destinations and deliver responses to the users. Anonymous proxy servers hide the user's IP address and other identifying information to provide basic anonymity. However, these servers are aware of both the source and the destination, and hence can trace user activities. Moreover, they have the weakest security against observers as monitoring in and out traffic of such a proxy server provides a high level of information about its users.

Figure 2 represents the geographic location of 1,441 public proxy servers obtained during Oct 11-17, 2010 from `proxy.org`, `publicproxyservers.com`, `proxy4free.com`, `freeproxy.ru`, and `tech-faq.com`. Note that, the figure is logarithmic scale. Among the available public proxy servers from 88 countries, most were located in the U.S. (i.e., 438) and in China (i.e., 250). Moreover, only 19 countries hosted more than 10 public proxy servers and 28 hosted a single server. These proxies were collected from major announcement lists and are a sample of available public proxy systems. Hence, this is not a complete list of public proxy servers but a representative sample.

In addition to volunteer-based systems, several *commercial anonymizer networks* such as `Anonymizer.com` and `GoTrusted.com` provide anonymous Internet access service to their clients. In these systems, clients pay a subscription fee to be able to relay their traffic through servers operated by the company. Usually, the user is connected to the network through a VPN tunnel and all traffic flows through the tunnel. However, as these companies are in charge of all the communications, they provide a lower degree of protection to their clients.

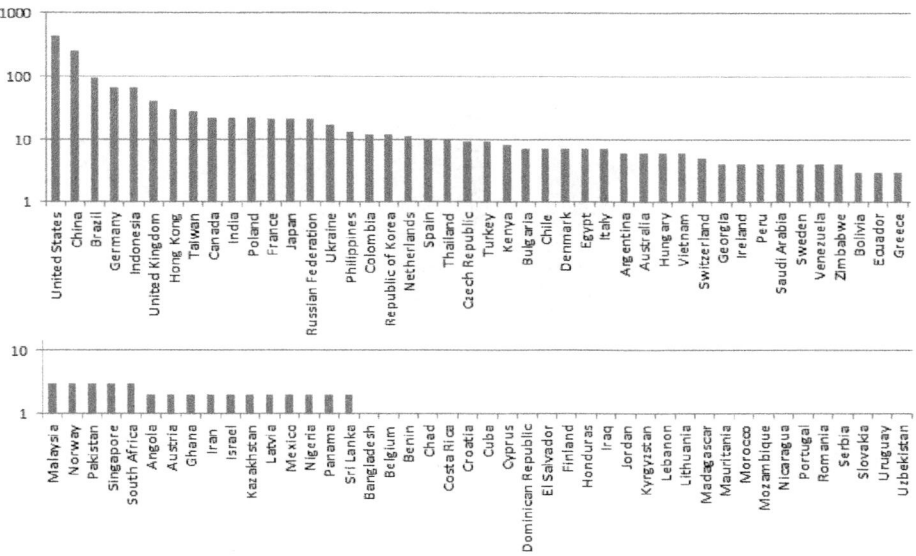

Fig. 2. Geographic Proxy Distribution (log-scale)

As proxy servers provide general web communication, *remailers* enable users to send electronic messages through their servers so that senders can not be traced. Remailers typically remove all identifying information from e-mails before forwarding them to their destination. Known examples of remailers include Cypherpunk, Mixmaster, and nym servers. However, due to heavy use of these servers by spammers in the past, they are not actively deployed any more. During our extensive web/blog search on Oct 2010, we were able to identify only 15 active remailers shown in Figure 3.

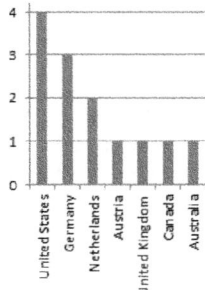

Fig. 3. Remailer Geo-Distributions

2.2   Mix Network

The building block of most of the current high-latency anonymity systems is the mix [4]. The basic building block of these systems, shown in Figure 4, is a set of mix processes where each mix process takes ciphertext messages that are encrypted with the mix process's public key as inputs. Mix process groups messages together as a batch and forwards the encrypted messages to the next mix process at certain flush times along with dummy messages.

Messages reach their destination after being forwarded by a set of mix processes through the network. For example in Figure 4, path $\mathcal{P}$ of a message M consists of 3 mix process Mix-1, Mix-2, and Mix-3. The client builds ciphertext C by encrypting message M along with random text R using each mix's public key K. The ciphertext (e.g., $E_1(A_{Mix-2}, R_1 + E_2(A_{Mix-3}, R_2 + E_3(D, R + M))))$) specifies the exact path the message will take through the mix network. Each mix node

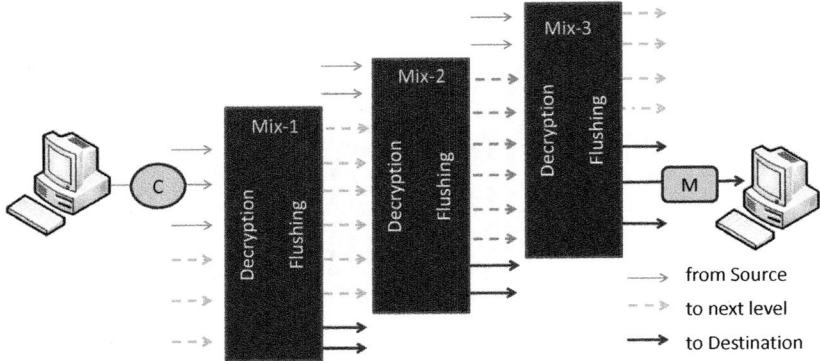

Fig. 4. A Mix Process

(e.g., Mix-1) receives the ciphertext decodes one layer to find next hop destination (e.g. A_{Mix-2}) and forwards payload (e.g., $E_2(A_{Mix-3}, R_2 + E_3(D, R+M)))$.

Asymmetric encryption and the flushing algorithms are the key for anonymity level and performance of a mix network. As encryption algorithms are provably secure with the current technology, flushing algorithms are an important component that may expose identity of the users. Flushing algorithms buffer incoming messages into a *pool* and forward messages in rounds. At each round, a random subset of the pool messages are mixed with dummy messages and flushed. The random subset can have a constant number or a dynamic number of messages. The duration of each round is decided based on a *threshold*. The threshold can be a number of messages N in the pool, or a timer counter T, or a combination of both.

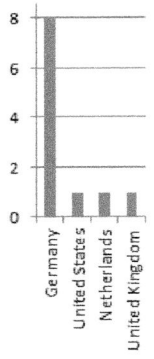

The *Java Anon Proxy* (JAP) is a mix network that uses servers provided by volunteers, usually institutions that declare conformance to JAP policies, to browse the Internet [2].

Fig. 5. JAP Geo-Distribution

JAP cascades encrypted packets through several mixes and keeps the traffic in a constant rate to avoid rate-based traffic analysis. The JAP program displays active mixes and users are able to select JAP cascades from those active mixes. Figure 5 presents the geographic location distribution of 11 JAP servers that were active on 12-19 Oct 2010. Compared to onion routing based systems Tor and I2P, JAP seems to have minimal usage at the time of our analysis.

2.3   Onion Routing

Onion routing is a low latency anonymous communication approach and is currently considered the most prevalent anonymization system design [10]. The basic idea of onion routing is similar to the mix system but performance is improved by using symmetric keys for relaying messages and asymmetric keys to establish circuits in the system.

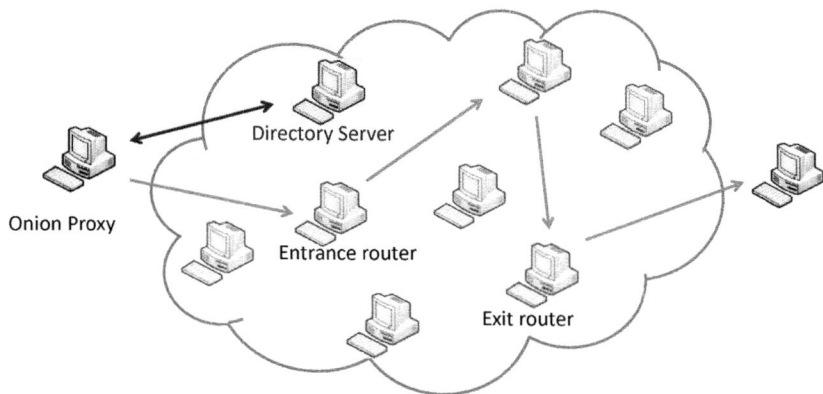

Fig. 6. The Onion Router (Tor) communication

There are different variations of onion routers such as Crowds [18], Tarzan [9], Invisible Internet Project (I2P) [1], and The Onion Router (Tor) [6] based on how the routing servers are organized; how the encryption algorithms are applied; how the tunnels are established; whether the transport-layer protocol uses TCP or UPD; or whether the clients relay traffic to other clients or not.

Tor, shown in Figure 6, is the most popular design as it combines the best parts of previous methods (e.g., the directory discovery of routing servers for clients, telescopic circuit establishment, and hiding locations). Directory servers are responsible for distributing signed information about known routers in the network [7]. Authoritative directory servers, currently 7 systems trusted by Tor developers [11], determine three-hop paths among volunteer servers using secured TCP connections. User messages are then encrypted as in mixes and forwarded through the established circuit to the dedicated exit router, which forwards the message to the final destination and echoes replies back. Entrance and exit nodes are particularly important as they know the source and the destination of the communication, respectively. Hence, authoritative directory servers pick only a subset of existing systems, which seems to be reliable, to become entry nodes and protect client profiling. Moreover, packets originate from the exit system from the destination's perspective and may be questioned regarding user actions. Hence, Tor allows relay systems to not become an exit node.

Figure 7 presents a snapshot of Tor servers based on their geographic location during Oct 20-24, 2010. For this analysis, we monitored the authoritative directory servers to determine the total number and geographical location of Tor servers. During the sampling period, we identified 10,510 unique servers at 95 countries but Tor system has approximately 1,500 active volunteers at a given time. Most of Tor relays are located in few countries. Similar to earlier studies [3, 14], Germany and U.S. had highest number of volunteers. Considering continents Europe had the highest number of servers. Interestingly, among Asian countries, Iran was third after technologically advanced countries such as Russia and Taiwan.

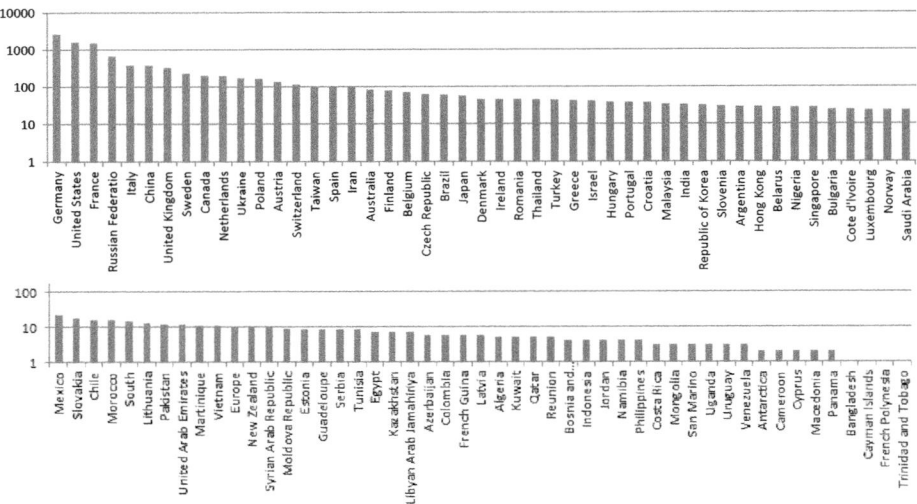

Fig. 7. Geographic Tor Server Distribution (log-scale)

Similar to Tor, the *Invisible Internet Project* (I2P) offers anonymization services that identity-sensitive applications can use. The I2P network is strictly message based, i.e., UDP, but there are libraries that allow reliable streaming communication on top of I2P network. Many applications can interact with I2P including mail, peer-to-peer, and IRC chat. Different from Tor, I2P does not focus on end-to-end delay and is preferred for peer-to-peer applications. To analyze its usage, we collected active I2P relays by joining the system during Oct 11-17, 2010. Figure 8 presents the geographic distribution of 483 servers in 29 countries (origin countries were determined by performing AS look-up of server IP addresses). Even though we had a longer sampling of I2P, we observed fewer servers than Tor system. Moreover, similar to Tor, Germany, U.S. and France had the highest number of volunteers.

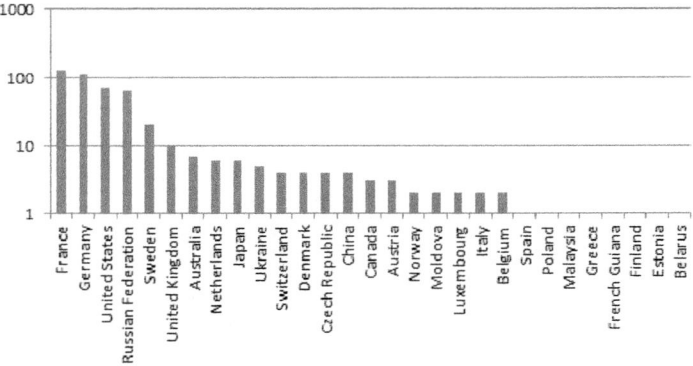

Fig. 8. Geographic I2P Server Distribution (log-scale)

3   Tor Usage Analysis

In this section, we analyze usage of Tor, currently the largest anonymity system. To be able to understand Tor network traffic, we set up two Tor relays using Tor 0.2.2.15- alpha. In order to analyze the traffic passing through our nodes, we used Wireshark to capture packet headers, i.e., IP addresses and port numbers for both source and destination, and payload size. During Oct 20-24, 2010, we had approximately 150 GB of data passing through our relays. According to the authoritative directory servers that provide bandwidth usage of each relay, our nodes were among the most popular relays of Tor in terms of bandwidth utilization.

Moreover, we inspected both incoming and outgoing traffic to observe whether our nodes were entry and exit routers. We observed client IPs when our relays were designated as entry nodes. Looking at IP addresses, we were able to identify the system we were communicating with. If the IP was not among Tor relay nodes, it either belonged to a user or to a server that users were communicating with. In order to distinguish between both, we looked at the payload size as Tor traffic is segmented into cells of 512 bytes. If the payload was 512 bytes that, the packet belonged to a user. Otherwise, the packet belonged to a server users were communicating with.

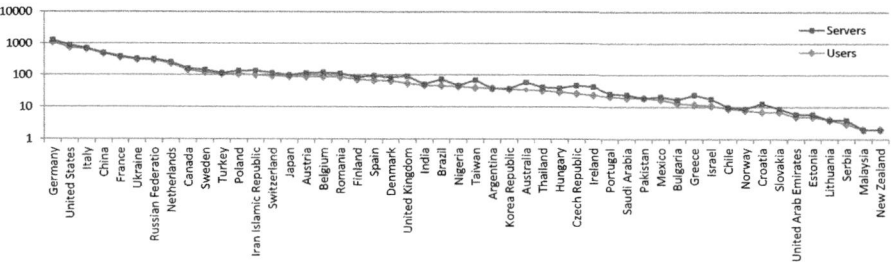

Fig. 9. Tor Usage (log-scale)

As part of our study, we also identified the geographical locations of clients and Tor relays. Table 1 and Figure 9 presents the number of Tor users and the relay servers from these countries. During a day period, when one of our servers was designated as an entry node, we observed **5,932** unique client IPs. According to the usage information we observed, Germany had the highest number of clients using Tor network and hosted most of the relays (similar to what was reported in [14]). Moreover, we analyzed the usage ratios of observed countries. For this, we

Table 1. Geographical distribution of Tor servers and clients

Country	Germany	U.S.	Italy	China	France	Russia	Netherlands	Canada	Sweden	Turkey
Users	1,076	734	657	469	356	289	223	143	119	108
Servers	205	141	42	29	32	27	29	18	25	6
Usage	5.48	.92	7.28	.36	2.64	1.60	5.01	.17	4.66	1.01

obtained the number of Internet users from `http://internetworldstats.com` and estimated the percentage of Tor usage in each country. Interestingly, Italy has the highest ratio of Tor usage relative to its Internet users.

During our data sampling, we also took snapshots of the authoritative directory servers to observe relay bandwidth. On average 1,567 Tor routers were observed to be active. Figure 10 presents the average contribution ratios of different countries in terms of total bandwidth, which was computed as the sum of all bandwidths of relays from a country.

Finally, to model the probability of each router forwarding a particular packet, we analyzed Tor relay usage from our nodes by counting the number of relay IPs. For an hour of traffic, we observed that 2% of relays carry 30% of traffic. Among the 15 most popular routers, 8 were in Germany, 4 in United States, 2 in France and 1 in Sweden. This indicates the disproportion of traffic carried by Tor servers and may weaken user anonymity [8].

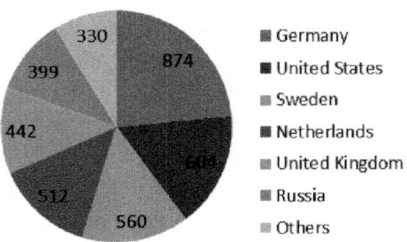

Fig. 10. Tor Bandwidth Distribuiton

4   Service Perspective

In this section, we investigate the usage of anonymity technology from a service perspective. These service applications include a secure web site at a university, spam emails, and peer-to-peer network. In total, 195,919 unique IP addresses were observed and analyzed to understand whether they originated from an anonymity system. For this, we compared the observed IP addresss to the collected IP addresses of anonymity servers in Section 2. Table 2 provides an overview of all the anonymity systems we looked at. The originating countries of these IP addresses were found using AS lookup.

Table 2. Analyzed Anonymity Systems

Network	Tor	I2P	JAP	Remailers	Proxies	Commercial
Servers	10,387	483	11	15	1,441	Anonymizer, GoTrusted
Service	General	peer-to-peer	General	E-mail	General	General

We collected the IP addresses of systems that accessed a *secure web site* from log files of more than 1 year. In this data, we had more than 21K unique IP addresses but there was no IP address from an anonymity server. This is expected because the secure web page requires login information and use of anonymizer would not improve anonymity of the user.

The following subsections provide our findings about spam e-mails and peer-to-peer traffic.

4.1   Spam Mail

Spam email data was collected using two approaches. First, we collected IP addresses of spam emails from Gmail accounts of coworkers and from departmental email servers during Oct 2010. Second, we gathered publicly available spam email IP addresses of recent spammers from the Internet. An important issue was to obtain recent data sets as anonymizer server IP addresses change over the time (except for commercial systems). As explained below, most spam e-mails were sent through relays in China and U.S. which is consistent with [16].

Gmail data set: We collected 4,843 IP addresses of spam e-mails from Gmail accounts of co-workers during Oct 2010. In this data, 42 IP addresses belonged to anonymity servers corresponding to 0.87 % of spam e-mails being sent through an anonymity network. Figure 11 presents the distribution of the anonymizer technology and the server geo-location. In this data set, I2P was utilized as spam relay more than the other sytems.

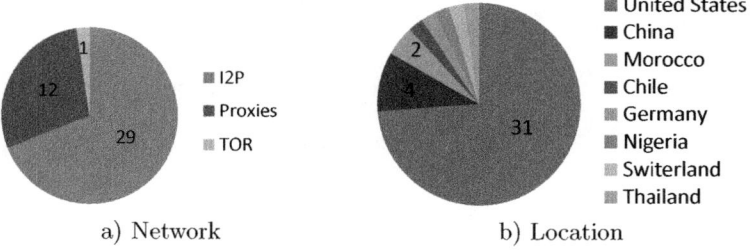

a) Network   b) Location

Fig. 11. Gmail Spam

Departmental data set: We collected 11,402 IP addresses of e-mails that were marked as spam by the departmental mail servers during Oct 2010. Among these IP addresses, only 76 were identified to arrive through an anonymity network corresponding to 0.67 % of total departmental spams. Figure 12 presents the distribution of utilized anonymizer technology and the server geo-location for departmental spam that was sent through an anonymity system. Similar to Gmail spam data, China and U.S. were the top two. In this data set, proxies and Tor network were utilized in sending spam e-mails.

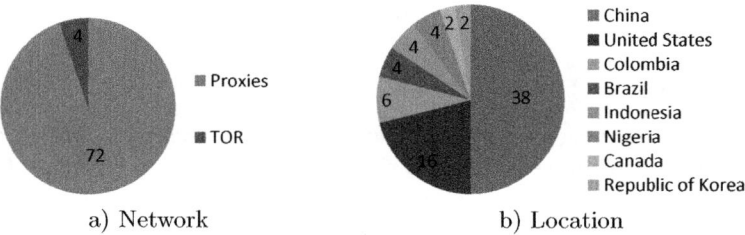

a) Network   b) Location

Fig. 12. Departmental Spam

Public data set: We collected 30,959 IP addresses that were recently marked as spam generators by public systems including `projecthoneypot.org`, `ipdeny.com`, `aclweb.org`, `landfall.net`, `spam-ip.com`, `spam-ip-list.blogspot.com`, and `spamlinks.net`. Among these IP addresses, 1,368 belonged to an anonymity server corresponding to 4.42 % of all spammer IPs. Figure 13 presents the distribution of utilized anonymizer technology and the server geo-location for spammer IP addresses in the data set. Similar to earlier data sets, China and U.S. were the two major relay nodes for spammers among the 31 countries observed and account for 65.4 % of all servers. In this data set, we observed that Proxy and Tor servers were utilized the most. Interestingly, the commercial anonmizer system `goTrusted.com` was utilized by spammers to send e-mails.

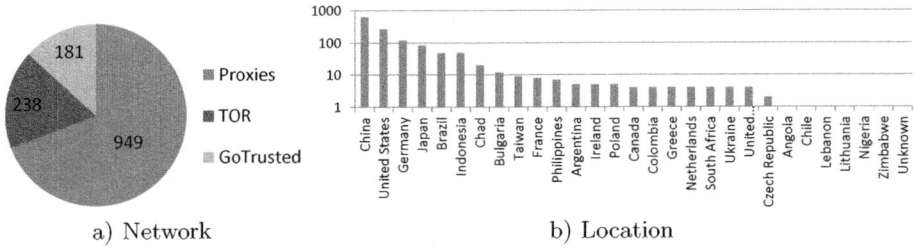

Fig. 13. Public Spam Data

All data sets: Figure 14 presents the results of all data combined (i.e., Gmail, department and public spam email data sets). Overall, proxies, `GoTrusted` and Tor were the three major sources utilized by spammers to relay e-mails. Moreover, servers in China, U.S. and Germany were the main relays of spam e-mails.

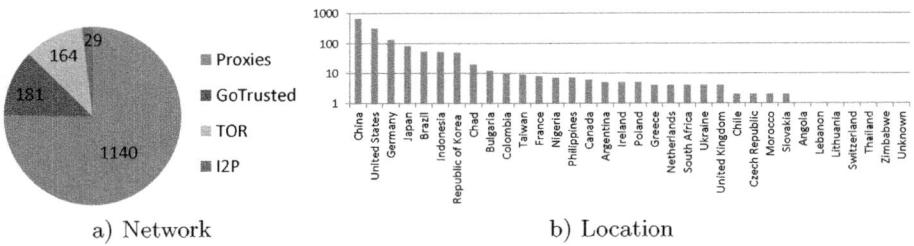

Fig. 14. Combined Spam Data

4.2   Peer-to-Peer Data

In order to analyze peer-to-peer traffic for anonymizer technology usage, we modified the open source Shareaza client, which joins BitTorrent, eDonkey, Gnutella, and Gnutella2 networks. The code was modified to log connected IP addresses and automatically search 3,600 keywords that were Google trends on Oct 2010 for about 50 countries. Considering copyright and other legal issues, the download feature was disabled so that no files were actually downloaded to our systems. We gathered data from two systems during Oct 10-24, 2010. In total, 114,593 unique IP addresses of peer-to-peer users were observed and analyzed.

Shareaza data set: Among the 114,593 IP addresses observed during our data collection, only 53 belonged to an anonymity system. Compared to the spam e-mail data set, this was very small. We believe that the main reason for this is the delay incurred by the anonymity system. Figure 15 presents the anonymity technology and geo-location distribution of the servers for the identified anonymizer relays. We observed that only Proxies and Tor servers were utilized by peer-to-peer clients to hide their IP addresses. Even though our peer-to-peer clients were in the U.S., only servers in Brazil, France, Hong Kong and Taiwan became relays to connect to our nodes. Among the 114,593 IP sources, United States and China accounted for most of them, but none of those utilized an anonymity network.

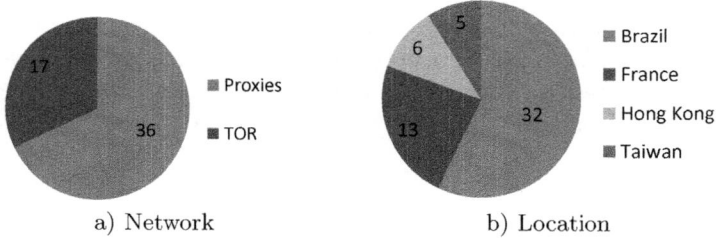

a) Network   b) Location

Fig. 15. Peer-to-peer data

Finally, within two weeks of data collection, we received a high ratio of bad queries among peer-to-peer client messages. These bad queries may be due to encrypted or compressed messages as reported by Chaabane et al. [3].

5   Related Work

There have been many studies on anonymity and anonymous systems and three studies have analyzed Tor usage as it gained popularity [3, 13, 14].

McCoy et al. looked for answers on how Tor is being used, how it is being mis-used, and who are its users [14]. In their experiments, the authors analyzed application-level protocols that use their nodes as exit node. According to their finding, interactive protocols, such as HTTP, make up 92 % of the connections and 58 % of bandwidth. Similarly, bit-torrent traffic consumes 40% of bandwith even though it accounts for 3.3 % of the connections. The authors also pointed to malicious usage of Tor routers and developed a method to detect malicious logging at exit routers. Moreover, they indicated that Tor has a global user base based on client distribution. Our results in Section 2 also indicate that Tor has the largest volunteer base among anonymity systems.

Moreover, Chaabane et al. performed a study to analyze applications that use Tor [3]. Authors monitored traffic on six servers which were pairwise located in U.S., Europe and Asia to inspect geo-diverse relays. Authors analyzed HTTP and BitTorrent traffic in detail. They pointed out that BitTorent consumes significant resources both in terms of packets and traffic size. Finally, authors pointed that Tor servers are used as 1-hop SOCKS proxies and present a method to detect such misuse.

Loesing et al. provided guidelines for a statistical analysis of Tor data focusing on countries of connecting clients and exiting traffic by port [13]. Pointing to privacy issues the authors derived guidelines for measuring sensitive data in anonymity networks. Moreover, they pointed to interesting cases such as increase in Tor usage by Iranian IP space in June 2009 after the Iranian elections; Tor blocking by China and consequent increase in bridge usage by Chinese IP addresses.

Our study is different from previous studies in that, in addition to Tor network analysis, we presented the analysis of other active anonymizer systems. We pointed out their usage and server geo-location distributions. Furthermore, we analyzed the traffic from different networks including a secure website, spam e-mails and peer-to-peer network. These studies allowed us to measure anonymizer usage in different domains.

6   Conclusion

Anonymity technologies have been utilized for a while. It is important to understand how people are using them, what applications are being used and which anonymity technology is popular. In this paper, we first summarized various anonymity technologies, i.e., proxy servers, mix networks and onion routing, and then focused on widely deployed anonymity systems, i.e., proxy servers, remailers, JAP, Tor, and I2P. For analyzing the current state of anonymizer networks, we joined them and collected information about relay nodes. We observed that similar countries, e.g., U.S., Germany and China, have the highest number of servers in different anonymizer networks.

Moreover, we set up Tor nodes as clients to collect entry and exit traffic information. Our servers relayed 150GB of data over five days. We observed that countries with high number of servers tend to have high number of Tor users. For instance, Germany and U.S. are top both in number of server and number of clients. Furthermore, to understand anonymity technology usage in different domains we analyzed spam emails and peer-to-peer clients. We observed that proxy servers were deployed more than other technologies. We believe that this is due to the higher latency in more secure systems.

Acknowledgments. This work was supported in part by National Institute of Justice.

References

1. I2p anonymous network, www.i2p2.de
2. Berthold, O., Federrath, H., Köpsell, S.: Web mixes: A system for anonymous and unobservable internet access. In: International Workshop on Designing Privacy Enhancing Technologies, pp. 115–129. Springer-Verlag New York, Inc., New York (2001)
3. Chaabane, A., Manils, P., Kaafar, M.: Digging into anonymous traffic: A deep analysis of the tor anonymizing network. In: 2010 4th International Conference on Network and System Security (NSS), pp. 167–174 (2010)
4. Chaum, D.L.: Untraceable electronic mail, return addresses, and digital pseudonyms. Commun. ACM 24(2), 84–90 (1981)
5. Danezis, G., Diaz, C.: A survey of anonymous communication channels (2008)
6. Dingledine, R., Mathewson, N., Syverson, P.: Tor: the second-generation onion router. In: SSYM 2004: Proceedings of the 13th Conference on USENIX Security Symposium, p. 21. USENIX Association, Berkeley (2004)
7. Edman, M., Yener, B.: On anonymity in an electronic society: A survey of anonymous communication systems. ACM Comput. Surv. 42(1), 1–35 (2009)
8. Feamster, N., Dingledine, R.: Location diversity in anonymity networks. In: Proceedings of the 2004 ACM Workshop on Privacy in the Electronic Society, WPES 2004, pp. 66–76. ACM, New York (2004)
9. Freedman, M.J., Morris, R.: Tarzan: a peer-to-peer anonymizing network layer. In: CCS 2002: Proceedings of the 9th ACM Conference on Computer and Communications Security, pp. 193–206. ACM, New York (2002)
10. Goldschlag, D., Reed, M., Syverson, P.: Onion routing. Communications of the ACM 42, 39–41 (1999)
11. Hahn, S., Loesin, K.: Privacy-preserving ways to estimate the number of tor users. Technical report, TOR project (November 2010)
12. Kelly, D.: A taxonomy for and analysis of anonymous communications networks. Technical report, Air Force Institute of Technology (March 2009)
13. Loesing, K., Murdoch, S.J., Dingledine, R.: A case study on measuring statistical data in the tor anonymity network. In: Workshop on Ethics in Computer Security Research (January 2010)
14. Mccoy, D., Kohno, T., Sicker, D.: Shining light in dark places: Understanding the tor network. In: Proceedings of the 8th Privacy Enhancing Technologies Symposium (2008)
15. Pfitzmann, A., Dresden, T., Hansen, M.: Anonymity, unlinkability, undetectability, unobservability, pseudonymity, and identity management – a consolidated proposal for terminology (2008)
16. Ramachandran, A., Feamster, N.: Understanding the network-level behavior of spammers. In: Proceedings of the 2006 Conference on Applications, Technologies, Architectures, and Protocols for Computer Communications, SIGCOMM 2006, pp. 291–302. ACM, New York (2006)
17. Reiter, M.K., Rubin, A.D.: Crowds: anonymity for web transactions. ACM Trans. Inf. Syst. Secur. 1(1), 66–92 (1998)
18. Reiter, M.K., Rubin, A.D.: Anonymous web transactions with crowds. Commun. ACM 42(2), 32–48 (1999)

Early Classification of Network Traffic through Multi-classification

Alberto Dainotti, Antonio Pescapé, and Carlo Sansone

Department of Computer Engineering and Systems, Universitá di Napoli Federico II
{alberto,pescape,carlosan}@unina.it

Abstract. In this work we present and evaluate different automated combination techniques for traffic classification. We consider six intelligent combination algorithms applied to both traditional and more recent traffic classification techniques using either packet content or statistical properties of flows. Preliminary results show that, when selecting complementary classifiers, some combination algorithms allow a further improvement – in terms of classification accuracy – over already well-performing stand-alone classification techniques. Moreover, our experiments show that the positive impact of combination is particularly significant when there are *early-classification* constraints, that is, when the classification of a flow must be obtained in its early stage (e.g. first 1 – 4 packets) in order to perform network operations online.

1 Introduction

Traffic Classification gained a lot of attention from both the industrial and academic research communities because of its application in several contexts: traffic/user profiling, network provisioning and resource allocation, QoS, enforcement of security policies, etc. While significant progress has been made in this field, with development in several research directions, literature clearly shows that there is still no *perfect* technique achieving 100% accuracy when applied to the entire traffic observed on a network link [20].

Deep Packet Inspection (DPI) is still considered the most accurate approach, but because of (i) computational complexity, (ii) privacy issues, and (iii) lack of robustness to the increasing usage of encryption and obfuscation techniques, it is used today as a reference (*ground-truth*) in order to evaluate the accuracy of new experimental algorithms that should overcome these limitations. Most of these algorithms are based on the application of machine-learning classification techniques to traffic properties and, even if their accuracy never reaches 100%, it has been shown that they typically are more resistant to obfuscation attempts and applicable when encryption is in place [5, 30].

In [13] we proposed the implementation in a single classification platform of combination strategies able to collect the results of very different classification techniques. Moreover, literature in the machine-learning field and pattern recognition [22] has produced several combination algorithms for building *multiclassifier* systems able to achieve better accuracy than each stand-alone classifier composing them.

J. Domingo-Pascual, Y. Shavitt, and S. Uhlig (Eds.): TMA 2011, LNCS 6613, pp. 122–135, 2011.
© Springer-Verlag Berlin Heidelberg 2011

In this work we apply several (6) of such algorithms to the problem of traffic classification, attempting the combination of classifiers (8) based on techniques known in the traffic classification field and we show preliminary results obtained from a real traffic trace. We show that in some cases it is possible to improve the overall classification accuracy over that of the best-performing classifier. Moreover, based on the observation that when a very limited quantity of information on each flow is available (which translates in less discriminating features) the accuracies of each stand-alone classifier decrease, we evaluate the improvement achieved by combining them under such conditions. Results show that the improvement is quite significant. This is important because several real-world applications of traffic classification as, for example, QoS, traffic shaping, and security policy enforcement, require *early-classification*, that is, the ability to generate a classification response when the flow is in its early stage (e.g. after 1 – 4 packets have been captured) [6] and thus could take real advantage from the use of the combination approaches here analyzed.

2   Related Work

A large amount of research work on traffic classification has been published in the past ten years, including several surveys and papers making comparisons among different techniques [20] [29] [10] [24]. All of them show *pros* and *cons* of different techniques and approaches as well as their inability to reach 100% classification accuracy. On the other side, research in the fields of machine-learning and pattern recognition has developed combination algorithms for classification problems that allow several improvements, included an increase in overall classification accuracy [22]. In the field of network traffic classification, a first rudimental combination approach to traffic classification was proposed in [26]: three different classification techniques are run in parallel (DPI, well-known ports and heuristic analysis), and a decision on the final classification response is taken only when there is a match between the results of two of them (otherwise the multi-classifier reports "unknown"). In [14] and [13], instead, we proposed the idea of combining multiple traffic classifiers using advanced combination strategies, inspired by research in the machine-learning and pattern-recognition fields related to multi-classification [22]. The approach of combining multiple classification techniques through specific algorithms to build a more accurate "multi-classifier", indeed, has been already used with success in other networking reasearch areas as network intrusion and anomaly detection [12]. As for traffic classification, concepts like *En-semble Learning* and *Co-training* have been introduced in [18], where a set of similar classifiers co-participate to learning, while an advanced combination of different traffic classification techniques has been shown in [9]. However, in that work, only variants of the *Dempster-Shafer* algorithm and a *majority vote* are taken into account, while in this paper we consider a more complete set of combination algorithms representative of the state of the art in multi-classification [22] – including those based on the Behavior Knowledge Space – plus we experiment on varying the composition of the pool of traffic classifiers.

Moreover, our contribution goes into a specific, and novel, direction by examining the impact of traffic classification under *early-classification* constraints. We pursue this target by evaluating the behavior of both the stand-alone classifiers and their combinations when trained and tested with discriminating features extracted only from a limited number of packets (from a single packet to the first ten packets). Several works have been presented that tackle the problem of early traffic classification [7] [11] [16], and they show the tradeoff between the amount of packets considered for extracting flow features and classification accuracy. In this work, for the first time we propose multi-classification as a way to improve accuracy while keeping the amount of information used for classification low.

3   Combination Algorithms

In many pattern recognition applications, achieving acceptable recognition rates is conditioned by the large pattern variability, whose distribution cannot be simply modeled. This affects the results at each stage of the recognition system so that, once it has been designed, its performance cannot be improved over a certain bound, despite the efforts in refining either the classification or the description method.

In the last years, some research groups concentrated the attention on a multiple classifier approach [8, 19, 21, 31]. The rationale of this approach lies in the assumption that, by suitably combining the results of a set of base classifiers, the obtained performance is better than that of any base classifier: it is claimed that the consensus of a set of classifiers may compensate for the weakness of a single classifier, while each classifier preserves its own strength [21]. The implementation of a multiple classifier system implies the definition of a *combiner* [22] for determining the most likely class a sample should be attributed to, considering the answers of the base classifiers.

Different combiners, independent of the adopted classification model, have been proposed in the literature [8, 22]. In the following we give a short introduction on the considered combiners. Since some traffic classifiers can be only seen as a *Type 1* classifier (i.e. a classifier that outputs just the most likely class), we considered only criteria that can be applied to classifiers that provide a crisp label as output. It is worth noting, in fact, that some well-known combination schemes (such as the *Decision Templates* proposed in [23]) cannot be applied to *Type 1* classifiers, since they require class probability outputs (i.e., the so-called *Type 3* classifiers[1]).

Before entering in details, it is worth recalling that some combiners make use of the so-called *confusion matrix* [31] for combining *Type 1* classifiers. The classification confusion matrix E^k is such that the generic element e_{ij}^k ($1 \leq i, j \leq m$, where m is the number of the classes) represents the percentage of samples belonging to the i-th class that the k-th classifier assigns to the j-th

---

[1] For the sake of completeness let us recall that *Type 2* classifiers operate at rank level, providing as output a subset of all the possible classes, with the alternatives ranked in order of plausibility of being the correct class.

class. Therefore, the value e_{ii}^k represents the percentage of samples belonging to the i-th class which are correctly classified by the k-th classifier. The values of the elements of E^k should be computed using a set of data (namely, a *validation set*) different from both the training and the test set.

1) Majority Voting (*MV*): each classifier votes for one class and the guess class is the one voted by the majority. If more classes obtain the same number of votes, the values e_{ii}^k are used for tie breaking, i.e. the vote of each classifier is weighted by the number representing the confidence degree of that classifier when it assigns a sample to the class it is voting for.

2) Weighted Majority Voting (*WMV*): in this case the confidence degree evaluated by means of the confusion matrices was used for weighting the votes given by each classifier. The combiner assigns each sample to the class C such that:

$$C = \arg\max_i \sum_k e_{ii}^k \cdot V_i^k \tag{1}$$

where V_i^k is 1 if the guess class of the k-th classifier is i and 0 otherwise.

3) Naïve Bayes (*NB*): the guess class is the one which maximizes the *a posteriori* probability. The probability that a sample belongs to the i-th class when the k-th classifier assigns it to the j-th class is assumed to be:

$$\frac{M_i \cdot e_{ij}^k}{\sum\limits_{h=1}^m M_h \cdot e_{hj}^k} \tag{2}$$

being M_i the number of samples belonging to the i-th class. Applying the Bayes' formula and standing the assumption of the independence of the classifiers, it can be simply shown, starting from the results presented in [22], that the class C which maximizes the *a posteriori* probability is:

$$C = \arg\max_i M_i \cdot \prod_{k=1}^N e_{ij}^k \tag{3}$$

where N is the number of classifiers and j is the guess class provided by the k-th classifier.

4) Dempster-Shafer combiner (*D-S*) [31]: this criterion is based on the Dempster-Shafer theory [17]. According to it, we define for each classifier, the *belief* in every possible subset A of the set $\Theta = \{A_1, A_2, \ldots, A_m\}$. In our context A_i is a proposition representing the fact that a sample is assigned to the i-th class by the considered classifier. The belief $bel(.)$ is calculated from a function, called *basic probability assignment*, which is denoted $m(.)$, by using the following equation:

$$bel(A) = \sum_{B \subseteq A} m(B) \tag{4}$$

where B is any subset of A. Obviously, we have $bel(A_i) = m(A_i)$ and $bel(\Theta) = 1$. In our case, when the k-th expert votes for the i-th class, we consider $m(A_i) = e_{ii}^k$

and $m(\Theta) = 1 - e_{ii}^k$. The values $m(A)$ supplied by each expert are combined via the Dempster rule, and the values $bel(A_i)$ are calculated using equation (4). The estimated class is the one that maximizes the value of $bel(A_i)$.

5) Behavior-Knowledge Space (*BKS*) method: one of the main drawbacks of the previously described approaches lies in the fact that they require (in a more or less explicit way) the independent assumption of the combining classifiers. This assumption does not usually hold in real applications, especially when the number of classifiers to be combined grows. More recently, a combiner has been proposed in order to overcome such limitations. It derives the information needed to combine the classifiers from a knowledge space, which can concurrently record the decision of all the classifiers on a suitable set of samples. This means that this space records the behavior of all the classifiers on this set, and then it is called the *Behavior-Knowledge Space* [19]. So, a Behavior-Knowledge Space is a N-dimensional space where each dimension corresponds to the decision of a classifier. Given a sample to be assigned to one of m possible classes, the ensemble of the classifiers can in theory provide m^N different decisions. Each one of these decisions constitutes one *unit* of the BKS. In the learning phase each BKS *unit* can record m different values c_i, one for each class. Given a suitably chosen data set, each sample x of this set is classified by all the classifiers and the *unit* that corresponds to the particular classifiers' decision (called *focal unit*) is activated. It records the actual class of x, say j, by adding one to the value of c_j. At the end of this phase, each *unit* can calculate the best representative class associated to it, defined as the class that exhibits the highest value of c_i. It corresponds to the most likely class, given a classifiers' decision that activates that *unit*. In the operating mode, the BKS acts as a look-up table. For each sample x to be classified, the N decisions of the classifiers are collected and the corresponding *focal unit* is selected. Then x is assigned to the best representative class associated to its *focal unit*.

6) Wernecke's (*WER*) method: it is similar to BKS and aims at reducing overtraining. The difference is that in constructing the BKS table, Wernecke [27] considers the 95 percent confidence intervals of the frequencies in each unit. If there is overlap between the intervals, the prevailing class is not considered dominating enough for labeling the unit. In this case, the "least wrong" classifier among the N members of the pool is identified, by using the confusion matrices. This classifier is authorized to assign the class to that unit. To calculate the 95 percent confidence intervals (CI), we used the Normal approximation of the Binomial distribution, as described in [22].

7) Oracle (*ORA*): when dealing with the evaluation of a MCS, it is useful to consider the performance of the so-called *"Oracle"*. The *Oracle* is the theoretic MCS that correctly classifies a sample if at least one of the base classifiers is able to provide the correct classification. It is evident that for a defined set of classifiers, the performance of the *Oracle* is an upper bound of all the MCS's obtainable from the same set of classifiers by using any combiner.

Table 1. Combination Algorithms

Label	Technique	Category	Training
NB	Naive Bayes	Bayesian	Confusion Matrix
MV	Majority Voting	Vote	Confusion Matrix
WMV	Weighted Majority Voting	Vote	Confusion Matrix
D-S	Dempster-Shafer	Dempster-Shafer	Confusion Matrix
BKS	BKS	Behavior Knowledge Space	BKS
WER	Wernecke	Behavior Knowledge Space	BKS&Confusion Matrix
ORA	Oracle	Oracle	na

4   The Tools Used

TIE[2] is a software platform for experimenting with and comparing traffic classification techniques. TIE allows the development of algorithms implementing different classification techniques as *classification plugins* (see Fig. 1) that are plugged into a unified framework, allowing their comparison and combination. We refer the reader to [13] as regards the TIE platform as well as the TIE-L7 classification plugin, which implements a DPI classifier using the techniques and signatures from the Linux L7-filter project [1] and that we used here to produce the ground truth. In the following, instead, we describe the new features we introduced in TIE in order to develop this work.

First of all, the above-mentioned combination strategies have been implemented in TIE's *decision combiner* (Fig. 1) and a set of support scripts have been developed in order to extract from the ground-truth (generated by TIE-L7) the confusion matrix and the BKS matrix needed for training the combiners. This information is reported into configuration files that are read at run time by the combiner selected by a command-line flag.

Moreover, in order to be able to rapidly test different machine-learning approaches to traffic classification we used the WEKA tool[3] that already implements a large number of machine-learning classification techniques. We plan to implement some of such techniques as TIE classification plugins, but in order to study and test a relevant number of machine-learning approaches we implemented a "bypass" mechanism in TIE which is structured in three phases:

- A new option allows, for each flow, to dump the corresponding classification features extracted by TIE (e.g. first ten packet sizes, flow duration, etc.) along with the ground truth label assigned by TIE-L7. Such information is dumped in a file in the *arff* format used by WEKA.
- The arff file is split in the *training* and *test sets* that are used to train and test various WEKA classifiers, whose classification output is in arff format too.
- A new TIE classification plugin is able to read the output of a WEKA classifier and use it to take the same classification decision for each flow. Multiple instances of such plugin can be loaded in order to support the output of several "WEKA" classifiers at the same time.

---

[2] http://tie.comics.unina.it

[3] http://www.cs.waikato.ac.nz/ml/weka

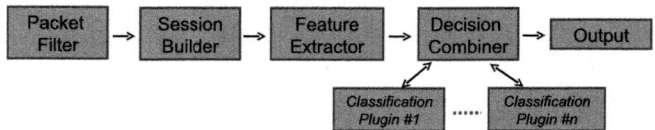

Fig. 1. TIE overall architecture

In this way TIE has a common view of both WEKA classifiers and TIE classification plugins: all these classifiers are seen as TIE plugins. This approach allowed us to easily test several classification approaches and to combine several of them plus pre-existing TIE classification plugins not based on machine-learning techniques (e.g. port-based and a novel lightweight payload inspection technique). In addition, based on the results of our studies on multi-classification we can later implement in TIE only the best performing classifiers.

Finally, in order to study the behavior of the classifiers and of the multi-classifier systems built on them, we introduced the option in TIE to generate a different file of features (in arff format) depending on the number of packets for each flow that can be used for extracting features. This option affects also the *native* TIE classification plugins that acquire the features directly by TIE's *feature extractor* (Fig. 1).

5   Data Set and Stand-Alone Classifiers

For the experimental results shown in this paper we used the traffic trace described in Table 2, in which we considered flows bidirectionally (*biflows* in the following) [13]. Each biflow has been labeled by running TIE with the TIE-L7 plugin in its default configuration, i.e. for each biflow a maximum of 10 packets and of 4096 bytes are examined.

Table 2. Details of the observed traffic trace

Site	Date	Size	Pkts	Biflows
Campus Network of the University of Napoli	Oct 3rd 2009	59 GB	80M	1M

From such dataset we then removed all the biflows labeled as UNKNOWN (about 167,000) and all the biflows that summed to less than 500 for their corresponding application label. Table 3 shows the traffic breakdown obtained[4]. This set was then split in three subsets in the following percentages:

- 20% classifiers *training set*
- 40% classifiers & combiners *validation set*
- 40% classifiers & combiners *test set*

---

[4] QQ is an instant messaging application.

Table 3. Traffic breakdown of the observed trace (after filtering out unknown biflows and applications with less than 500 biflows)

Application	Percentage of biflows
BITTORRENT	12.76
SMTP	0.78
SKYPE2SKYPE	43.86
POP	0.24
HTTP	16.3
SOULSEEK	1.06
NBNS	0.14
QQ	0.2
DNS	4.08
SSL	0.21
RTP	1.16
EDONKEY	19.21

Table 4. Stand-alone classifiers

Label	Technique	Category	Features
J48	J48 Decision Tree	Machine Learning	PS, IPT
K-NN	K-Nearest Neighbor	Machine Learning	PS, IPT
R-TR	Random Tree	Machine Learning	L4 Protocol, Biflow duration & size, PS & IPT statistics
RIP	Ripper	Machine Learning	L4 Protocol, Biflow duration & size, PS & IPT statistics
MLP	Multi Layer Perceptron	Machine Learning	PS
NBAY	Naive Bayes	Machine Learning	PS
PL	PortLoad	Payload Inspection	Payload
PORT	Port	Port	Ports

We considered eight different traffic classifiers, summarized in Table 4. The first five are based on Machine-Learning approaches common in literature both in terms of algorithms and discriminating features [28, 3, 25, 4]. As regards the features, in the same table *PS* stands for Payload Size, while *IPT* means Inter-Packet Time [15]. The J-48, K-NN, MLP, and NBAY classifiers consider the vectors of the first 10 PS and IPT, whereas the R-TR and RIP classifiers use statistics of PS and IPT as their average and standard deviation. The latter classifiers also take into account the transport-level protocol of the biflow, the biflow duration (in milliseconds) and size (in bytes). The *PortLoad* classifier, instead, is a light-weight payload inspection approach, recently presented in [2], that overcomes some of the problems of DPI, as computational complexity and invasiveness, at the expense of a reduced accuracy. PortLoad only uses the first 32 bytes of transport-level payload from the first packet (carrying payload) seen in each direction. Finally, we also considered the traditional traffic classifier based on transport-level protocol ports.

Table 5 shows the classification accuracy (i.e. percentage of correctly classified biflows) of every stand-alone classifier for each application and over the entire test set. The different performance of the classifiers for every application, and in particular the best accuracy score for each of them (printed in bold font), show that they have some complementarities. Moreover, the *Port* classifier has a very low overall score, which in general would suggest to avoid its use in a multi-classifier system, but we considered it because it reaches very high accuracy values for some specific applications. Finally, the last column contains the accuracies that would be obtained by the oracle, that is, by selecting for

Table 5. Classification accuracy – per-application and overall – of stand-alone classifiers (best values are in bold font) and oracle

Class	J48	K-NN	R-TR	RIP	MLP	NBAY	PL	PORT	ORACLE
					Classifier				
Bittorrent	98.8	97.4	**98.9**	98.6	55.1	79.9	7.7	21.0	99.9
SMTP	95.1	92.9	93.8	**96.0**	90.6	69.2	8.2	96.3	99.4
Skype2Skype	98.8	97.2	96.5	**99.2**	94.6	31.8	98.7	0	99.7
POP	96.0	95.0	98.7	93.9	0	79.6	29.2	**100**	100
HTTP	99.5	98.9	**99.6**	99.3	94.3	63.3	99.1	47.7	100
Soulseek	**98.6**	96.8	98.3	98.1	93	97.7	0	0	99.9
NBNS	78.4	75.9	79.9	**80.4**	9	0	0	0	85.4
QQ	0	0.7	**2.5**	0	0	0	0	0	3.2
DNS	93.6	92.6	95.3	94.4	51.1	86.2	**100**	99.7	100
SSL	96.1	93.1	95.2	93.7	69.5	68.2	**99.1**	0	99.6
RTP	**84.0**	74.1	64.5	77.3	0	41.5	0	0	92.2
EDonkey	93.0	91.7	**93.3**	91.5	72	16.1	92.9	0.1	95.7
overall	**97.2**	95.9	96.3	97.0	82.3	43.7	83.7	15.6	98.8

each biflow the correct response when this is given by at least one of the stand-alone classifiers. The overall accuracy obtained by the oracle (98.8%) shows that the combination of these classifiers can theoretically bring an improvement with respect to the best standalone classifier (97.2%).

6   Experimental Evaluation of Combiners

We experimented the combination of the stand-alone classifiers from the previous section using the algorithms explained in Section 4. When combining the classifiers we experimented with different pools of them, as shown in Table 6, where the overall accuracies for each pool and combiner are reported. The values show that in general it is indeed possible to gather an improvement through combination, as suggested theoretically by the oracle, but this improvement depends not only on the combiner adopted, but also on the choice of the classifiers. The port-based classifier has in general a negative impact on the performance of the multi-classifier system, the same happens for the Naive Bayes classifier. This behavior can be easily explained by looking at their rather low performance as stand-alone classifiers (Table 5). In particular, the performance of the MV and the WMV combiners dramatically depend on the weak performance of the Naive Bayes classifier, since the worst pools for these combiners are those in which this classifier is present. This can be explained by considering that the worst performance of the Naive Bayes classifier happen on the classes in which also the MLP classifier performs quite bad, so lowering the performance of voting-based combiners. On the contrary, the D-S combiner and the multinomial approach followed the BKS methods and are able to cope with such a situation. Finally, since the independent assumption of the base classifier does not always hold, the Naive Bayes combiner does not perform very well on the average.

The pool of classifiers achieving the best results is reported in Table 6 in bold fonts, using 6 classifiers out of the 8 tested, and closely followed by the second

Table 6. Classification accuracy of each combiner for different pools of classifiers combined (best or close to the best values are reported in bold font)

Pool of classifiers								Combiner					
J48	K-NN	R-TR	RIP	MLP	NBAY	PL	PORT	NB	MV	WMV	D-S	BKS	WER
X		X	X					54.1	96.3	96.3	96.2	**97.7**	**97.7**
X		X	X			X		55.2	96.4	96.2	96.6	**97.8**	**97.8**
X	X	X	X	X				53.5	90.7	90.7	**96.7**	96.0	96.1
X	X	X	X	X	X			80.1	72.0	72.2	96.7	**97.3**	**97.3**
X	**X**	**X**	**X**	**X**		**X**		93.5	90.8	91.0	97.0	**97.9**	**97.9**
X	X	X	X	X	X	X		80.9	72.0	72.2	97.0	**97.7**	**97.7**
X	X	X	X	X		X	X	93.6	90.5	90.8	97.1	**97.7**	**97.7**
X	X	X	X	X	X	X	X	54.6	72.8	71.2	97.1	**97.4**	**97.4**

pool in the table that includes only 4 classifiers. As for the combiners, the same table shows that the best accuracies (percentages in bold fonts) are achieved by the combiners based on the Behavior Knowledge Space (BKS and Wernecke), with the highest score of 97.9% overall accuracy. This value should be interpreted by considering the highest overall accuracy achieved by a stand-alone classifier (97.2%) and the maximum theoretically possible combination improvement set by the oracle (98.8%): an improvement equal to 43% of the maximum achievable.

We then focused our experiments on the context of early classification. This subject has been previously investigated in literature because of its important applicative characteristics, being early classification indispensable to perform on-line classification of traffic flows: a new traffic flow is observed on a link and the system must identify as soon as possible the application associated to it (e.g. in order to apply a security policy to the flow). In such case, therefore, classification cannot be performed with all the flow information available, and literature [6, 2] has shown that there is indeed a trade-off between the ability of classifying a flow using only its first packets and the classification accuracy. In our experimental analysis we investigated the benefits of multi-classification in this context. We therefore repeated all the training and testing of the stand-alone classifiers previously considered with a variable amount of information available, that is by varying the number of packets for each flow from which the discriminant features were extracted. We also repeated the combination experiments varying the number of packets and considering the *J48,R-TR,RIP,PL* pool of classifiers. We chose this pool because its overall accuracy values are very close to those of the best pool but a reduced number of classifiers is used. Moreover, all the classifiers from this pool use algorithms with a small computational complexity. This is particularly relevant in the context of online classification.

Table 7 shows the performance of the stand-alone classifiers when 1 to 10 packets are used to extract classification features. The PortLoad classifier is based on a technique that uses at most 2 packets, therefore accuracies related to more than 2 packets are all equal, whereas the port-based classifier needs a single packet to perform classification. The best accuracy value for each number of available packets is reported in bold font. The results in the table confirm the impact of reducing the amount of available information on classification accuracy, as suggested by the literature. Moreover, the values in this table can

Table 7. Classification accuracy of each stand-alone classifier depending on the number of packets used for the feature extraction (the highest accuracy for each column is reported in bold font)

Classifier	Number of packets observed for each biflow									
	1	2	3	4	5	6	7	8	9	10
J48	62.1	**94.6**	**95.9**	96.0	**96.8**	**97.1**	**97.2**	**97.2**	**97.2**	**97.2**
K-NN	62.4	91.5	92.8	95.0	94.9	94.9	95.4	95.7	95.6	95.9
R-TR	72.7	93.4	93.6	94.9	95.3	96.8	96.0	96.0	96.1	96.2
RIP	69.5	93.7	94.7	**96.2**	96.1	96.5	96.7	96.9	96.9	96.9
MLP	43.5	71.7	81.0	82.3	82.3	82.3	82.3	82.3	82.3	82.3
NBAY	31.5	39.9	42.6	43.7	43.7	43.7	43.7	43.7	43.7	43.7
PL	**76.2**	83.7	83.7	83.7	83.7	83.7	83.7	83.7	83.7	83.7
PORT	15.6	15.6	15.6	15.6	15.6	15.6	15.6	15.6	15.6	15.6

Table 8. Classification accuracy when varying the number of available packets. Pool of combined classifiers: *J48,R-TR,RIP,PL.*

Combination	Number of packets observed for each biflow									
	1	2	3	4	5	6	7	8	9	10
MV	57.8	93.9	94.4	95.6	95.9	96.2	96.3	96.3	96.4	96.4
D-S	83.1	96.0	96.9	97.0	97.4	97.4	96.4	96.5	96.5	96.5
BKS	97.0	98.4	98.3	98.4	98.4	98.4	98.4	98.4	98.4	98.4
WER	97.0	98.3	98.2	98.4	98.4	98.4	98.4	98.4	98.4	98.4

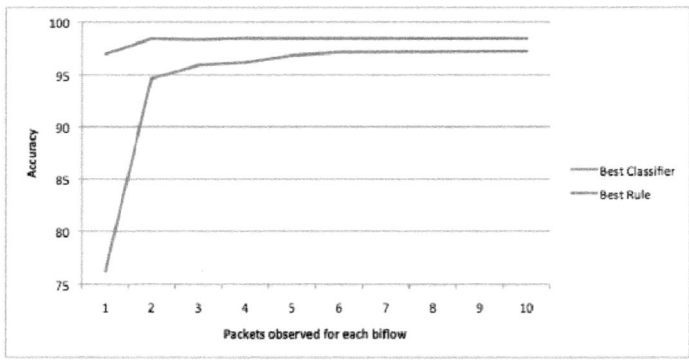

Fig. 2. Classification accuracy of the best-performing stand-alone classifier (blue line) *vs* the multi-classifier (red line)

be compared with the results of multi-classification reported in Table 8. Here, to reduce the large amount of experiments, we limited our tests to only four combiners (including the best two methods). The overall accuracy values show that in the case of early-classification the impact of multi-classification is rather significant, this is also visible in Figure 2 where we plotted for each number of available packets both the highest accuracy achieved by stand-alone classifiers (blue line) and the highest accuracy achieved by the combiners (red line): for 1 packet the combination brings an improvement of about +21% overall accuracy, while for 2 packets it is of about +4%. Such large improvements suggest that multi-classification may be an effective strategy for the implementation of more accurate traffic classifiers able to work online in the context of early classification.

7 Conclusion

In this work we have presented and evaluated different combination techniques for traffic classification, including the BKS-based algorithms, which were not previously proposed in the traffic classification field. Moreover, for the first time we proposed the use of multi-classification in the context of early traffic classification. The preliminary experimental results here presented show several findings:

- The combination of stand-alone classifiers that present complementarities can improve the overall classification accuracy.
- The combiners based on the Behavior Knowledge Space look more promising than the others with respect to traffic classification. This behavior can be due to the fact that in our case the independent assumption of the combining classifiers does not hold. Moreover, the availability of a significant amount of training data does not cause BKS overtraining (which is one of the main drawbacks of this method).
- Even if literature has shown that the transport-level port is still a useful classification feature, combiners cannot effectively exploit the (small) discriminating power of a port-based traffic classifier. On the contrary, classifiers based on (light-weight) payload inspection complement very well with machine-learning classifiers.
- The positive impact on overall accuracy of combination is particularly significant in the context of early classification. With very strict requirements (e.g. 1 or maximum 2 packets per biflow) the performance decrease in terms of classification accuracy of the stand-alone classifiers can be almost entirely compensated by their combination.

As future work, we plan to extend this investigation to traffic traces from different links. Moreover we will focus furthermore on the exploitation of multi-classification in the context of early-classification, investigating in detail also computational complexity and timing-related issues. For this purpose we will also implement the machine-learning classifiers that were best performing as TIE plugins.

Acknowledgments

The research activities presented in this paper have been partially funded by Accanto Systems and by LATINO project of the FARO programme jointly financed by the Compagnia di San Paolo and by the Polo delle Scienze e delle Tecnologie of the University of Napoli Federico II. The authors would like to thank Antonio Quintavalle for his support to experimental activities.

References

1. L7-filter, Application Layer Packet Classifier for Linux,
   `http://l7-filter.sourceforge.net`
2. Aceto, G., Dainotti, A., de Donato, W., Pescapé, A.: PortLoad: taking the best of two worlds in traffic classification. In: IEEE INFOCOM 2010 - WiP Track (March 2010)
3. Alshammari, R., Zincir-Heywood, A.N.: Machine learning based encrypted traffic classification: identifying ssh and skype. In: CISDA 2009: Proceedings of the Second IEEE International Conference on Computational Intelligence for Security and Defense Applications, pp. 289–296. IEEE Press, Piscataway (2009)
4. Auld, T., Moore, A.W., Gull, S.F.: Bayesian neural networks for internet traffic classification. IEEE Transactions on Neural Networks 18(1), 223–239 (2007)
5. Bernaille, L., Teixeira, R.: Early recognition of encrypted applications. In: Uhlig, S., Papagiannaki, K., Bonaventure, O. (eds.) PAM 2007. LNCS, vol. 4427, pp. 165–175. Springer, Heidelberg (2007)
6. Bernaille, L., Teixeira, R., Akodjenou, I., Soule, A., Salamatian, K.: Traffic classification on the fly. ACM SIGCOMM CCR 36(2), 23–26 (2006)
7. Bernaille, L., Teixeira, R., Salamatian, K.: Early Application Identification. In: ACM CoNEXT (December 2006)
8. Bloch, I.: Information combination operators for data fusion: a comparative review. IEEE Trans. System Man and Cybernetics, Part A 26(1), 52–76 (1996)
9. Callado, A., Kelner, J., Sadok, D., Kamienski, C.A., Fernandes, S.: Better network traffic identification through the independent combination of techniques. Journal of Network and Computer Applications 33(4), 433–446 (2010)
10. Callado, A., Szabó, C.K.G., Gero, B.P., Kelner, J., Fernandes, S., Sadok, D.: A Survey on Internet Traffic Identification. IEEE Communications Surveys & Tutorials 11(3) (July 2009)
11. Carela-Español, V., Barlet-Ros, P., Solé-Simó, M., Dainotti, A., de Donato, W., Pescapé, A.: K-dimensional trees for continuous traffic classification. In: Ricciato, F., Mellia, M., Biersack, E. (eds.) TMA 2010. LNCS, vol. 6003, pp. 141–154. Springer, Heidelberg (2010)
12. Corona, I., Giacinto, G., Mazzariello, C., Roli, F., Sansone, C.: Information fusion for computer security: State of the art and open issues. Information Fusion 10(4), 274–284 (2009)
13. Dainotti, A., de Donato, W., Pescapé, A.: Tie: A community-oriented traffic classification platform. In: Papadopouli, M., Owezarski, P., Pras, A. (eds.) TMA 2009. LNCS, vol. 5537, pp. 64–74. Springer, Heidelberg (2009)
14. Dainotti, A., de Donato, W., Pescapè, A., Ventre, G.: Tie: A community-oriented traffic classification platform. In: Technical Report TR-DIS-102008-TIE, Dipartimento di Informatica e Sistemistica, Universitá degli Studi di Napoli Federico II (October 2008)
15. Dainotti, A., Pescapè, A., Ventre, G.: A packet-level characterization of network traffic. In: CAMAD, pp. 38–45. IEEE, Los Alamitos (2006)
16. Gómez Sena, G., Belzarena, P.: Early traffic classification using support vector machines. In: LANC 2009: Proceedings of the 5th International Latin American Networking Conference, pp. 60–66. ACM, New York (2009)
17. Gordon, J., Shortliffe, E.: The dempster-shafer theory of evidence. In: Buchanan, B.G., Shortliffe, E. (eds.) Rule-Based Expert Systems, pp. 272–292. Addison-Wesley, Reading (1984)

18. He, H., Che, C., Ma, F., Zhang, J., Luo, X.: Traffic classification using en-semble learning and co-training. In: AIC 2008: Proceedings of the 8th Conference on Applied Informatics and Communications, pp. 458–463. World Scientific and Engineering Academy and Society (WSEAS), Stevens Point (2008)
19. Huang, Y.S., Suen, C.Y.: A method of combining multiple experts for the recognition of unconstrained handwritten numerals. IEEE Trans. Pattern Analysis and Machine Intelligence 17(1), 90–94 (1995)
20. Kim, H., Claffy, K., Fomenkov, M., Barman, D., Faloutsos, M., Lee, K.: Internet traffic classification demystified: myths, caveats, and the best practices. In: CoNEXT 2008: Proceedings of the 2008 ACM CoNEXT Conference, pp. 1–12. ACM, New York (2008)
21. Kittler, J., Hatef, M., Duin, R.P.W., Matas, J.: On combining classifiers. IEEE Trans. Pattern Analysis and Machine Intelligence 20(2), 226–239 (1998)
22. Kuncheva, L.I.: Combining Pattern Classifiers: Methods and Algorithms. Wiley-Interscience, Hoboken (2004)
23. Kuncheva, L.I., Bezdek, J.C., Duin, R.P.W.: Decision templates for multiple classifier fusion: an experimental comparison. Pattern Recognition 34(2), 299–314 (2001)
24. Nguyen, T.T., Armitage, G.: A Survey of Techniques for Internet Traffic Classification using Machine Learning. IEEE Communications Surveys and Tutorials (2008) (to appear)
25. Park, J., Tyan, H.R., Kuo, C.C.J.: Ga-based internet traffic classification technique for qos provisioning. In: International Conference on Intelligent Information Hiding and Multimedia Signal Processing, pp. 251–254 (2006)
26. Szabo, G., Szabo, I., Orincsay, D.: Accurate traffic classification, pp. 1–8 (June 2007)
27. Wernecke, K.D.: A coupling procedure for discrimination of mixed data. Biometrics 48, 497–506 (1992)
28. Williams, N., Zander, S., Armitage, G.: Evaluating machine learning algorithms for automated network application identification. Tech. Rep. 060401B, CAIA, Swinburne Univ. (April 2006)
29. Williams, N., Zander, S., Armitage, G.: A preliminary performance comparison of five machine learning algorithms for practical ip traffic flow classification. ACM SIGCOMM CCR 36(5), 7–15 (2006)
30. Wright, C.V., Monrose, F., Masson, G.M.: On inferring application protocol behaviors in encrypted network traffic. Journal of Machine Learning Research 7, 2745–2769 (2006)
31. Xu, L., Krzyzak, A., Suen, C.Y.: Method of combining multiple classifiers and their application to handwritten numeral recognition. IEEE Trans. Syst. Man Cybernetics 22(3), 418–435 (1992)

Software Architecture for a Lightweight Payload Signature-Based Traffic Classification System

Jun-Sang Park, Sung-Ho Yoon, and Myung-Sup Kim

Dept. of Computer and Information Science, Korea University, Korea
{junsang_park,sungho_yoon,tmskim}@korea.ac.kr

Abstract. Traffic classification is a preliminary and essential step for achieving stable network service provision and efficient network resource management. While a number of classification methods have been introduced in the literature, the payload signature-based classification method shows the highest performance in terms of accuracy, completeness, and practicality. However, the payload signature-based method has a significant drawback in high-speed network environments; the processing speed is much slower than that of other classification methods such as the header-based and statistical methods. In this paper, we describe various design options to improve the processing speed of traffic classification in designing a payload signature-based classification system, and we describe choices we made for designing our traffic classification system. Also, the feasibility of our design choices was proved via experimental evaluation on our campus traffic trace.

Keywords: payload-signature, traffic classification, flow analysis.

1 Introduction

As individual and corporate users are becoming increasingly dependent on the Internet, network speeds are increasing and a variety of services and applications are being developed. Thus, there is a growing need for monitoring and analyzing Internet traffic from the application perspective for achieving efficient network operation and management in various areas, for example, pay-for billing, CRM, SLA, etc. Further, the need for traffic monitoring and analysis will continue to increase. Effective methods are needed for analyzing many types of application-level traffic and handling real-time processing for the large amounts of traffic on high-speed links.

Traffic classification is a preliminary and essential step for achieving stable network service provision and efficient network resource management. While a number of classification methods have been introduced in the literature, the payload signature-based classification method shows the highest performance in terms of accuracy, completeness, and practicality. [1, 3, 4, 8, 9] However, the processing speed of the current classification system is insufficient for real-time handling of the large amounts of traffic on high-speed networks.[6, 7, 11] Given the increasing number of applications and greater usage of applications that generate large amounts of traffic, the inadequate processing speed of payload-based analysis is a challenge that must be mitigated. In this paper, we will define the factors affecting the processing speed of

J. Domingo-Pascual, Y. Shavitt, and S. Uhlig (Eds.): TMA 2011, LNCS 6613, pp. 136–149, 2011.
© Springer-Verlag Berlin Heidelberg 2011

the signature-based classification system. We aim to improve the processing speed by proposing an optimal classification system structure based on the experimental evaluation of possible design choices for the classification system.

This paper is organized as follows. Section 2 describes research related to this issue and Section 3 describes the design considerations needed for the current payload-based classification systems. Section 4 presents the factors affecting processing speed. An optimal solution based on the experimental results is suggested in order to improve the processing speed. In Section 5, the proposed method is applied to our classification system and its validity is proven. Finally, in Section 6, conclusions and future research directions are described.

2 Related Works

Many applications try to bypass the firewall for a seamless service by frequently changing traffic patterns, so the signature appears in a complex form. In addition, due to the increase of network-based applications and application layer protocols, the number of signatures necessary for identifying applications has been increasing. As the number of signatures and their complexity increase, the processing speed of the payload signature-based classification system has become an important element in determining the performance of the traffic classification system. Many ongoing studies on payload signature-based classification systems aim to accelerate the classification process, but most studies have focused on improving the performance of the pattern-matching algorithm.

Table 1. Comparison of two classification systems

Tool	Signature Format	No. of Signatures	bps	Matching Algo.
L7-filter	Regular Expression	About 70	Less than 10Mbps	NFA
Snort	Explicit String + Regular Expression	About 5000	Less than 100Mbps	DFA

Table 1 compares the classification speed of two popular traffic analysis systems, Snort and Linux L7-filter. The table shows the configuration of the signature-based classification system. The L7-filter is widely used for application-level traffic classification. It uses regular expressions to represent signatures and NFA (Non-deterministic Finite Automata) for pattern-matching. However, when over 70 signatures are used to classify applications, it shows a processing speed of 3.5Mbps or less. [6] DFA (Deterministic Finite Automaton) has been proposed to increase the processing speed of NFA. Snort applies the DFA-based pattern matching method, but it has a processing speed of less than 100Mbps [6, 7, 11]. These two systems have tended to focus on pattern matching algorithms to improve performance. However, the time complexity of the matching algorithm is wholly dependent on the configuration of the input data, resulting in limited performance improvement. Thus, real-time traffic analysis of high-speed links (Gbps) might be insufficient if they are

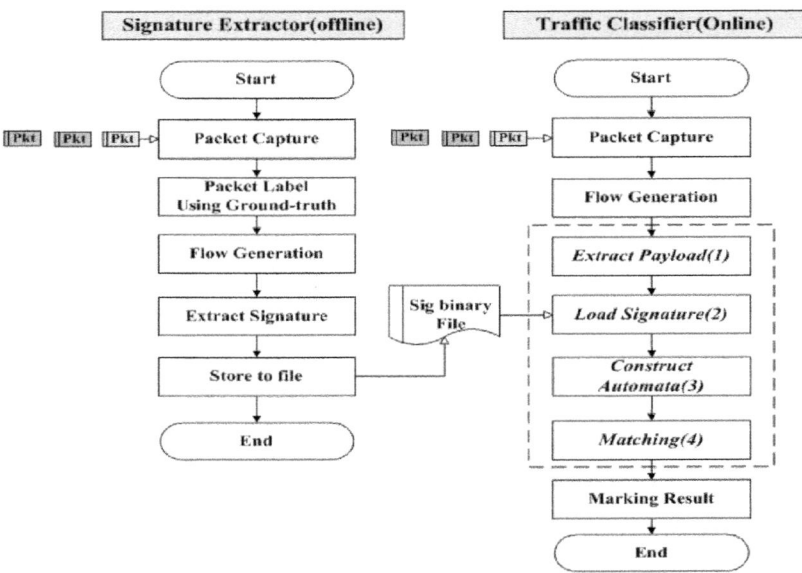

Fig. 1. Flow Chart of our Traffic Classification System

only based on the elaboration of the matching algorithm. Thus, we must consider other options to improve the processing seed of the traffic classification system.

Figure 1 illustrates a payload-based traffic classification system in the form of a flow chart, which consists of two main modules: the payload signature extraction module and traffic classification module with the extracted signatures. We have developed this system and deployed it in our campus network for real-time classification of campus Internet traffic. A total of 845 payload signatures were extracted via the payload-signature extraction system. The system specification was Intel ® Core2 Duo E7200 2.53GHz CPU with 3GByte memory. The average processing speed of our system was 160Mbps of Internet traffic. This speed is insufficient to support the link rate (up to 300Mbps) of our campus Internet traffic.

The most time-consuming aspect of the classification system was the signature matching module in Figure 1. This is shown in Table 2, which compares the execution times of modules 1, 2, 3, and 4 in our classification system. The signature matching module was the most time consuming and used about 83% of the total processing time. In this paper, we propose a means to utilize a matching algorithm and show how to optimize the search space in order to improve the processing speed of the payload signature-based traffic classification system.

Table 2. Comparison of module execution time ratios

Module	Load Pkt Payload	Load Signature	Construct Automata	Matching
Execution Time Ratio	0.31%	7.63%	8.72%	83.33%

3 Considerations for Performance Enhancement

In this section, we describe three considerations needed to improve the processing speed of the payload signature-based traffic classification system. The current classification method described in the previous section is not optimized for these three considerations. The three considerations include the matching algorithm, input data search space, and signature search space.

First, we need to select the best matching algorithm and signature representation method for traffic classification. There is no matching algorithm that is optimized for all types of input data. Table 3 describes the time complexities of several string pattern matching algorithms used in various fields, including traffic analysis. We categorized their signature types into two groups: an explicit string and a regular expression. Each algorithm provides a different level of performance based on the features of its input data. For example, the processing speed of the Boyer-Moore algorithm increases when there is more frequent matching of string characters with strings in the payload. However, because of the nature of the payload, it is likely that available characters will be present and repeated characters will be infrequent. Thus, we cannot guarantee performance. There is the additional problem that we cannot guarantee the performance of DFA or NFA, because their performance depends on the frequency of the wildcard character * in the signature string.

Table 3. Time–complexity comparison of several matching algorithms

Algorithm		Preprocessing time	Matching time
Explicit String	Brute-force	No preprocessing	$\Theta(nm)$
	DFA	$\Theta(m \, |\Sigma|)$	$\Theta(n)$
	Robin-Karp	$\Theta(m)$	$\Theta((n-m+1) \, m)$
	Boyer-Moore	$\Theta(m + |\Sigma|)$	$\Omega(n/m), O(n)$
Regular Expression	NFA	$\Theta(m \, |\Sigma|)$	$\Omega(n^2)$
	DFA	$\Theta(m \, |\Sigma|)$	$\Omega(1)$

n: Payload length, m: Signature length , Σ: A number of available characters

Second, we need to optimize the search space in the flow of the selected matching algorithm. The flow-based analysis, other than of the packet-based one, is popularly used in traffic analysis. In addition, we need to minimize the number of packets in a flow and limit the byte size in the packet that will be searched. Figure 2 shows the distribution of the matched location of signatures when we applied our classification system with about 845 signatures. Most signatures were found within the first two packets in a flow and within the first 500 bytes in the packets, as is shown in Figure 2. However, our classification system performs matching for every packet in the flow and all bytes in a packet, which results in significant performance degradation. We believe that we can utilize this experimental result to reduce the search space of the input data for the matching algorithm.

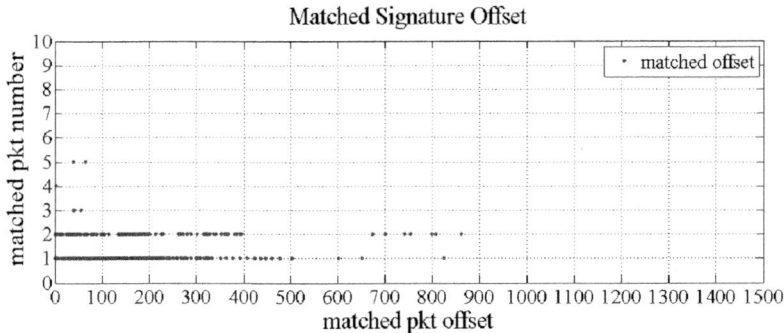

Fig. 2. Matched Signature Offset

Third, we need to consider how the order of signature matching depends on the change of traffic conditions. Our current classification system does not reflect the change of traffic patterns over time. By operating our system on our campus network, we learned that the signature locality can be found over time. Aggressive use of only a small number of signatures out of the total number can identify the traffic flows during a certain time period. Further, the active signatures change over time.

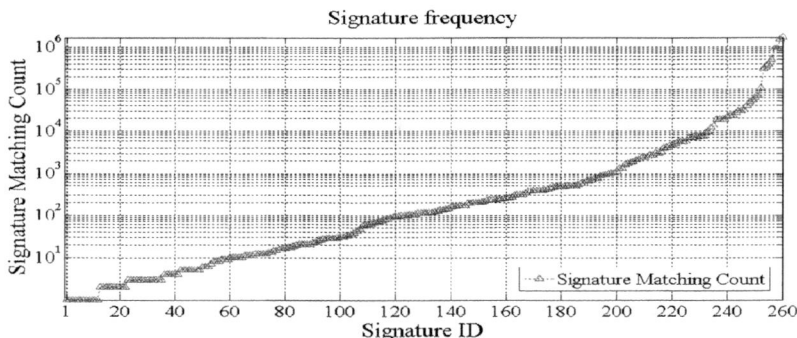

Fig. 3. Matched Signature Hit Count

Figure 3 shows the cumulative hit count for traffic signatures, based on the existing classification system, over a period of five hours. Hits occurred for a total of 260 signatures out of 845. Only about 60 signatures had hit counts greater than 1000. Therefore, the number of applications available at a certain time is small and most of the traffics can be classified by using only a few signatures for the classification system.

4   Classifier Optimizations

In this section, we propose our solutions for the problems mentioned in Section 3. We are aiming to optimize the processing speed of the classification system. We will prove the validity of these solutions via experiments.

4.1 Matching Algorithm and Signature Representation

In this section, we will describe the attempt to optimize our classification system in terms of the matching algorithm and signature representation. The performance of the matching algorithm varies depending on how the signatures are represented. Therefore, it is reasonable to select many different matching algorithms based on various signature types. Also, we will describe the signature format we selected in order to optimize the classification algorithm.

Table 4. Comparison of the matching algorithm performance according to signature types

Matching Algorithms / Signature Type		Robin-Karp	DFA Full	NFA Partial	NFA Full
Explicit String	Fixed offset	**0.03 sec**	0.05 sec	0.08 sec	0.08 sec
	Variable offset	1.28 sec	**0.32 sec**	0.90 sec	0.42 sec
Regular Expression	".*" <= 2	3.45 sec	0.19 sec	**0.08 sec**	0.16 sec
	".*" > 3	1.35 sec	0.06 sec	**0.05 sec**	0.55 sec

Table 4 shows the processing time of four popular matching algorithms for various signature types. We initially divided signature types into two categories: explicit string and regular expression, based on the method of signature representation. An explicit string is further classified into fixed offset and variable offset according to the signature location in the input data. A regular expression is classified into two types according to the frequency of the wildcard character '*' in a signature, as is shown in Table 4.

In order to determine the best matching algorithm according to the type of signature representation, we divided the signatures into 4 types and applied each signature type to the Robin-Karp, DFA and NFA matching algorithms. The experimental result shows that the performance of the matching algorithms varies depending on the signature types. Thus, there is no single best matching algorithm for all input data types. The algorithm performance can vary depending on the features of each input data type.

Table 5. Selected matching algorithms according to signature types

Signature Type		Matching Algorithm	Library
Explicit String	Fixed offset	Robin-Karp	Self-implemented
	Variable offset	DFA Full	Boost library
Regular Expression	".*" <= 2	NFA Partial	PCRE library
	".*" > 3	NFA Partial	PCRE library

Table 5 shows our selection of matching algorithm. We selected different matching algorithms in accordance with the signature types based on our experiment. For a signature in the form of an explicit string with a fixed offset, the Robin-Karp string matching algorithm is used. For an explicit string signature with a variable offset, the DFA full matching algorithm is selected. For signatures in the form of regular

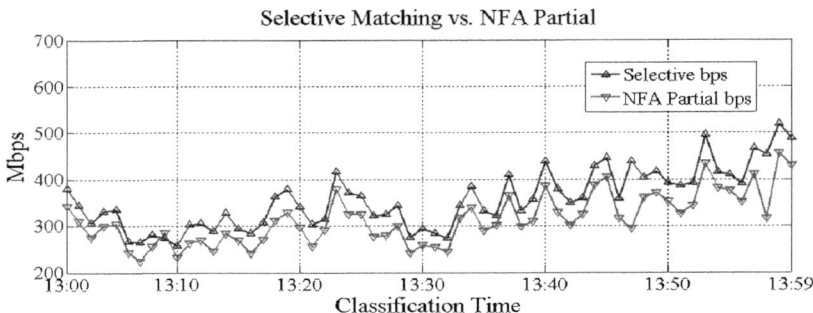

Fig. 4. Performance of Selective Matching Method versus NFA Partial Method

expressions, the NFA Partial matching method is selected. For the experiments, we implemented the Robin-Karp algorithm ourselves, and we utilized the Boost and PCRE libraries for the DFA and NFA algorithms, respectively.

Figure 4 shows the performance of our selective matching method in comparison with the NFA-Partial method that gives the highest average performance for a single matching algorithm for all types of signatures. The y-axis in Figure 4 indicates the maximum throughput (bps) of the matching method. The proposed selective method can handle an average of 50Mbps more than that of the NFA-Partial method.

Previously, the signature was mostly represented as an explicit string, but it is more often represented as a regular expression nowadays. Considering signatures in the form of regular expressions, the frequency of the characters '^', '$', '.', '*', "[]" is a factor affecting the processing speed of the matching algorithm. By using '^' and '$', the matching algorithm can improve the processing speed. The use of the wildcard character '*' and bracket expression "[]" causes an increase in the number of states for an automata. So, the processing speed of the matching algorithms can be increased by using wildcards and brackets. We utilized this information to refine all of our 845 signatures.

Table 6. Signature distribution before and after refinement

Feature	Before refinement	After refinement
Signatures with '^' or '$'	45.45%	50.78%
Signatures with more than 3 ".*"	6.49%	3.48%
Signatures with "[]"	0%	0%

Table 6 represents the signature proportions before and after refining the signatures. The percentage of signatures where '^' and '$' are present was increased, and the percentage of expressions with more than three "*" characters was decreased. In addition, no signature expressions used the bracket expression "[]".

The classification rate has been improved by about 1-2 seconds after refining the signatures. This improvement is due to the reduction in the use of wildcards and increase in the incidence of expressions using a signature in fixed offset form with '^' and '$' characters.

4.2 Optimization of Search Space

This section describes how to optimize the search space of the input data and the signature provided to the classification system.

4.2.1 Inspected Packet Count in Terms of Flow and Packet Size

The processing time of the classification system can be reduced by limiting the number of packets inspected in the flow. Table 7 shows the measurements of performance as the number of packets inspected in a flow increases. The packets inspected are defined as the first n packets with a payload after the TCP connection setup. According to the analysis results shown in Table 7, the classification accuracy and completeness increases as the number of packets inspected increases, but they are almost identical after the fifth packet. This is because most connections send a low number of control packets before sending the contents that are common among all of the same connection types. Most payload signatures are extracted from the first several packets in a flow. Therefore, the classification result is sufficiently accurate and the classification time is reduced by limiting the number of packets inspected to the first 5 packets in each flow.

In case of inspection of the first packet in a flow, completeness is less than 80% because the classification system could not classify mfile[14] flows. The application flows identified the second packet in the flow.

A limit on bytes inspected in a packet needs to be considered in order to reduce the search space of the input data of the classification system. In accordance with the experimental result shown in Figure 2, we limited the inspected byte size in a packet to the first 1000 bytes. We can reduce the processing time of our classification system by limiting the number of packets inspected in a flow and limiting the byte size in a packet. This allows our system to cope with more bandwidth from a high-speed link.

Table 7. Performance measurement according to the number of packets inspected

No. of pkts inspected / Performance		Pkt1	Pkt2	Pkt3	Pkt4	Pkt5	Pkt6	Pkt7
Completeness (%)	Flow	92.3	93.1	93.1	93.2	93.2	93.2	93.2
	Packet	82.4	85.2	86.5	86.7	86.8	86.8	86.8
	Byte	77.5	81.1	82.2	82.5	82.7	82.7	82.7
Accuracy (%)	Flow	96.2	96.3	96.3	96.3	96.3	96.3	96.3
	Packet	98.6	98.7	98.7	98.7	98.7	98.7	98.7
	Byte	97.4	97.6	97.6	97.6	97.6	97.6	97.6

4.2.2 Two-Level Hierarchical Signature Structure

We can reduce processing time by constructing the payload signatures in the form of a 2-level hierarchical structure, which consists of an application protocol-level signature and an application-level signature, as is shown in Figure 5. Nowadays, a number of different applications commonly use an application-level protocol for various purposes. For example, the HTTP protocol is widely used by many applications. It is reasonable to distinguish these types of HTTP traffic according to their applications rather than considering only a single type of HTTP traffic. In our signature hierarchy, HTTP traffic is detected at the first level, and the application name is then determined at the second level. The signature hierarchy is defined by an inclusion relationship. If all traffic identified by using signature S_X can be classified by using signature S_Y, then S_Y includes S_X. S_Y is called an application protocol-level signature, while S_X is called an application-level signature.

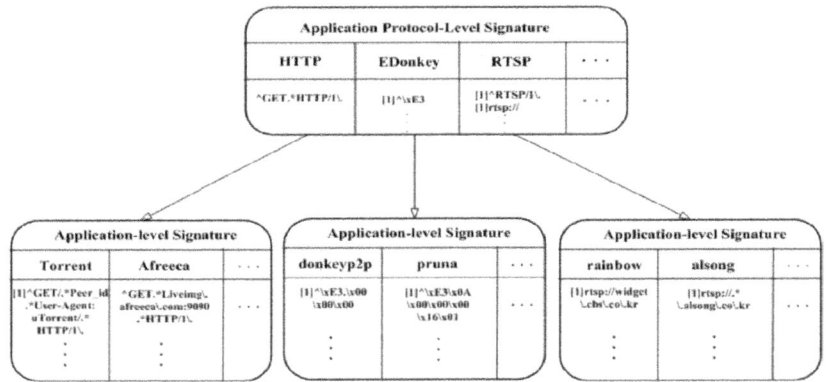

Fig. 5. Two-level Hierarchical Signature Structure

The classification system identifies input flow via the application protocol-level signature. If a flow was classified by using an application protocol-level signature, then the classification system can identify flow via the application-level signatures included in the application protocol-level signature. This analysis reduces the signature search space of the classification system and reduces processing time.

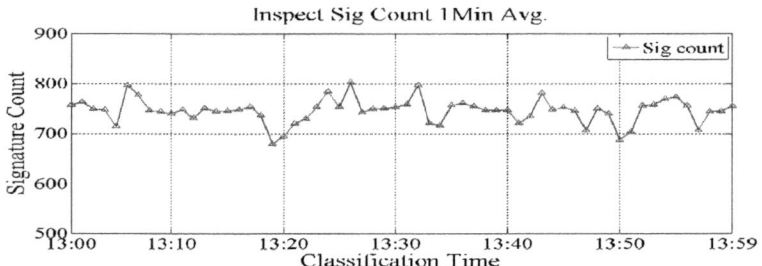

Fig. 6. Inspected Signature Count per Flow

Figure 6 shows the average number of signatures required to analyze a single flow. Our current analysis method compares every signature in order to analyze a flow, while the suggested analysis method based on the hierarchical structure can improve the processing speed of the classification system by reducing the search space. At this stage, we can achieve a reduction of more than 100 signatures out of 845 signatures.

4.2.3 Multiple Signatures with Single Automaton

As the number of signatures increases, the signature search space increases and matching becomes more time-consuming. One method to reduce the number of signatures is to use a single automaton for multiple signatures. However, analysis time can be reduced only if the matching time of a single automaton for a group of signatures is shorter than the sum of the matching times of multiple automatons for a group of signatures. Table 8 shows the analysis time of DFA and NFA, before and after grouping signatures, for NateOn messenger, a widely used application in Korea.

Table 8. Matching time

	Signature	DFA Full	NFA Partial
	^GET.*NateOn.*	0.23 secs	0.05 secs
	^GET.*nateon\.nate\.com.*	0.22 secs	0.06 secs
SSSA	^GET.*adimg\.nate\.com.*	0.18 secs	0.11 secs
	^GET.*cyad\.nate\.com.*	0.25 secs	0.15 secs
	^GET.*nateonipml\.nate\.com.*	0.21 secs	0.10 secs
MSSA	^GET.*(NateOn)/(nateon\.nate\.com)/(adimg\.nate\.com) /(cyad\.nate\.com)/(nateonipml\.nate\.com).*	0.22 secs	9.24 secs

MRSA(Multiple Signatures Single String) is a signature consisting of a group of 5 SSSA(Single Signature Single Automaton) signatures. DFA spends almost the same amount of time on classification before and after grouping, while the classification time for NFA sharply increases after grouping. Table 8 suggests that we can shorten classification time by classifying signatures after grouping in the form of MSSA.

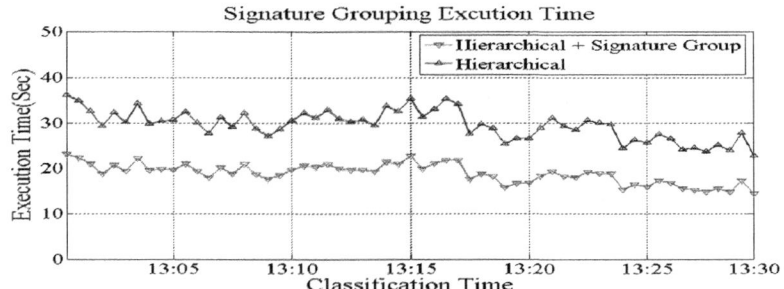

Fig. 7. Comparison of Processing Speed Before and After Signature Grouping

Figure 7 shows the improvement of processing speed achieved by signature grouping. The search space was also decreased because grouping reduces the number of signatures. The figure shows a reduction of 20 seconds after grouping.

4.2.4 Signature Locality for Optimization

As shown in Figure 3, when the number of applications running on the target network is limited and only some particular applications are running during a certain time period, the relevant application traffic exhibits regional characteristics. Therefore, most traffic can be classified by using a few signatures during that specific time period. Thus, we

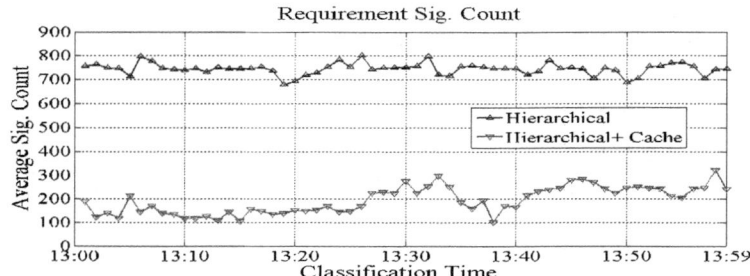

Fig. 8. Inspected Signature Count per Flow

can minimize the search space by first examining frequently occurring signatures by dynamically changing the signature memory structure according to the signature hit ratio. We call this signature caching.

Figure 8 describes the average number of signatures used to identify a flow with and without the proposed signature caching method. We used the exponential average value of the signature hit count to update the signature cache structure. By constantly updating the hit count via the exponential average, the system can deal with changes in application usage according to the time flow.

The proposed signature caching method improved the processing speed of the classification system by reducing the signature search space by 20%. This analysis method based on signature frequency can minimize the signature search space because it examines the signature with the highest current hit count first.

Cache hits occurred for a total of 211 signatures out of 845. More than 70% of the flows had hit only 20 signatures.

5   Experimental Evaluation

In this section, we apply the proposed method to traffic data collected from a real campus network, and we then prove the validity of the method.

5.1 Traffic Trace

In this section, we describe how a traffic classification and verification system is used to validate our suggested method, by applying it to a campus network.

Figure 9 is a diagram that shows the verification method for the location of traffic collection, configuration of the classification system, and verification of the classified traffic. All packets are collected on the link connecting the campus network to the Internet via TCS. The collected flows are transferred to TAS and classified by the application unit based on the payload signature.

In order to verify the accuracy of the results, ground-truth traffic information is collected via TMA[2,5]. TMA is installed on the terminal host and creates information based on the socket, including the process name, IP, port, protocol, and path. After checking the open socket on the regular host on which TMA is installed, the TMA information is sent to TMS, which integrates the TMA information from

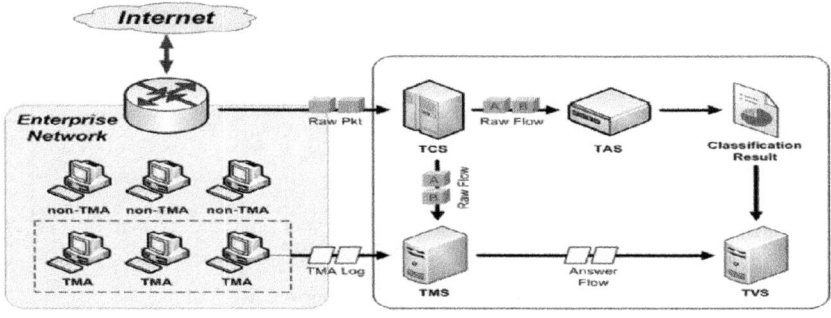

TCS : Traffic Capture System   TAS : Traffic Analysis System   TVS : Traffic Verification System
TMA : Traffic Measurement Client TMS : Traffic Measurement Server

Fig. 9. Configuration of Classification and Verification Network

Table 9. Summary of Traffic Trace Used in Experiment

Algorithm	TCP	UDP	Total
Flow	3,972,069	2,462,886	6,434,955
Packet(K)	218,013	113,094	331,107
Byte(MB)	188,415	100,332	288,747

each host and provides the ground-truth information for the results of the classification system. The performance evaluation is done at TVS by comparing the result of TAS with the ground-truth information.

Table 9 shows the results of the traffic trace used in the experiment. Traffic was collected at the Internet connection point of our campus network, and it was comprised of traffic from a variety of applications used by approximately 3000 hosts.

5.1.1 Configuration of Proposed Classification System

Table 10 presents the classification methods we applied to optimize the speed of our classification system, based on the experimental results presented in Section 4.

Table 10. Summary of proposed method

Coverage	Process : 260 Application : 126 Signature : 845
Classification Criteria	Application(Set of Processes)
Classification Unit	Bidirectional Flow
Inspected Packet Offset	1st~5th packet in flow, 1000 bytes in a packet
Matching Algorithm	Selective(RK, DFA-Full, NFA-Partial)
Automata	MSSA

5.2 Evaluation and Analysis

Figure 10 shows the results of the signature-based analysis used to evaluate the performance of the proposed methods. We compare our proposed method with the previous system. The graph shows the time spent on classification. The results were

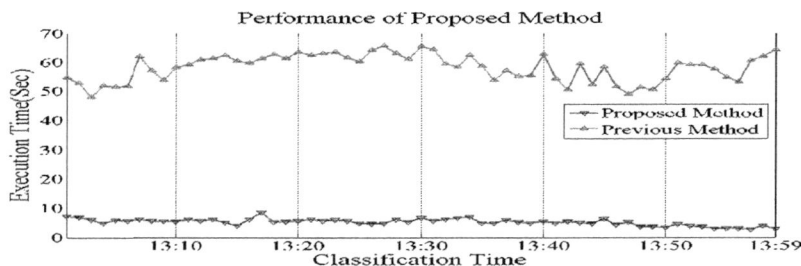

Fig. 10. Performance of the Proposed Method

obtained by proactively checking signatures in order to minimize the search space. Compared to the previous system, a 5-fold improvement in processing speeds can be seen. The traffic trace has different amounts of traffic, but the proposed method has an absolute classification time, because the signature search space is not affected by the amount of traffic.

Table 11 presents the classification result of the previous and proposed methods. Our signature-based classification system achieved more than 95% accuracy and 80% completeness. We can reduce the search space of classification system, but the classification result is not affected.

Table 11. Accuracy & Completeness of Proposed Method

Metric Method	Accuracy			Completeness		
	Flow	Packet	Byte	Flow	Packet	Byte
Previous Method	95.57	98.31	98.70	91.17	84.25	81.21
Proposed Method	95.57	98.31	98.70	91.17	84.25	81.21

6   Conclusion and Future Work

In this paper, we optimized the factors that affect the processing speed of a payload signature-based traffic classification system. We experimentally evaluated each factor and suggested a method for creating an efficient classification system. The suggested method showed a 5-fold improvement in processing speed over our previous processing classification system.

This method provides a software-based means to improve the processing speed of general classification systems in the proposed computing environment. We plan to design a hardware-based classification system that will allow real-time analysis on a large-scale network.

Acknowledgments

This work was supported by Basic Science Research Program through the National Research Foundation of Korea (NRF) funded by the Ministry of Education, Science and Technology (2009-0090455).

References

1. Park, J.-S., Park, J.-W., Yoon, S.-H., Oh, Y.-S., Kim, M.-S.: Development of signature Generation system and verification network for application-level traffic classification. In: Conference of Korea Information Processing Society, Busan, April 23-24, vol. 16(1), pp. 1288–1291 (2009)

2. Yoon, S.-H., Roh, H.-G., Kim, M.-S.: Internet Application Traffic Classification using Traffic Measurement Agent. In: Conference of Korea Information Communication Society, Jeju, July 2-4, p. 618 (2008)

3. Sen, S., Spatscheck, O., Wang, D.: Accurate, scalable in-network identification of p2p traffic using application signatures. In: World Wide Web, New York, USA, May 17-20 (2004)

4. Risso, F., Baldi, M., Morandi, O., Baldini, A., Monclus, P.: Lightweight, Payload-Based Traffic Classification An Experimental Evaluation. In: IEEE International Conference on Communications, Beijing, China, May 19-23, pp. 5869–5875 (2008)

5. Yoon, S.-H., Park, J.-W., Oh, Y.-S., Park, J.-S., Kim, M.-S.: Internet Application Traffic Classification Using Fixed IP-port. In: APNOMS, Jeju, Korea, September 23-25, pp. 21–30 (2009)

6. Yu, F., Chen, Z., Dino, Y., Lakshman, T.V., Katz, R.H.: Fast and memory Efficient Regular Expression Matching for Deep Packet Inspection. In: ANCS, San jose California USA (2006)

7. Hayes, C.L., Luo, Y.: DPICO: a high speed deep packet inspection engine using compact finite automata. In: ACM/IEEE Symposium on Architecture for Networking and Communications Systems, Orlando, Florida, USA, December 03-04 (2007)

8. Liu, H., Feng, W., Huang, Y., Li, X.: Accurate Traffic Classification. In: Networking, Architecture, and Storage, NAS (2007)

9. Park, B.-C., Won, Y.J., Choi, M.-J., Kim, M.-S., Hong, J.W.: Empirical Analysis of Application-Level Traffic Classification Using Supervised Machine Learning. In: Ma, Y., Choi, D., Ata, S. (eds.) APNOMS 2008. LNCS, vol. 5297, pp. 474–477. Springer, Heidelberg (2008)

10. Vasiliadis, G., Polychronakis, M., Antonatos, S., Markatos, E.P., Ioannidis, S.: Regular expression matching on graphics hardware for intrusion detection. In: Kirda, E., Jha, S., Balzarotti, D. (eds.) RAID 2009. LNCS, vol. 5758, pp. 265–283. Springer, Heidelberg (2009)

11. Cormen, T.H., Leiserson, C.E., Rivest, R.L., Stein, C.: String Matching. In: Introduction to Algorithms, 2nd edn., ch. 32, pp. 906–932. MIT Press, McGraw-Hill (2001)

12. Holzer, M., Kutrib, M.: Descriptional and computational complexity of finite automata. In: Dediu, A.H., Ionescu, A.M., Martín-Vide, C. (eds.) LATA 2009. LNCS, vol. 5457, pp. 23–42. Springer, Heidelberg (2009)

13. Lovis, C., Baud, R.: Fast exact string pattern-matching algorithms adapted to the characteristics of the medical language. J. Am. Med. Inform. Assoc. 7, 378–391 (2000)

14. mfile (as of January 2011), http://mfile.co.kr

Mining Unclassified Traffic Using Automatic Clustering Techniques

Alessandro Finamore, Marco Mellia, and Michela Meo

Politecnico di Torino
lastname@tlc.polito.it

Abstract. In this paper we present a fully unsupervised algorithm to identify classes of traffic inside an aggregate. The algorithm leverages on the K-means clustering algorithm, augmented with a mechanism to automatically determine the number of traffic clusters. The signatures used for clustering are statistical representations of the application layer protocols.

The proposed technique is extensively tested considering UDP traffic traces collected from operative networks. Performance tests show that it can clusterize the traffic in few tens of pure clusters, achieving an accuracy above 95%. Results are promising and suggest that the proposed approach might effectively be used for automatic traffic monitoring, e.g., to identify the birth of new applications and protocols, or the presence of anomalous or unexpected traffic.

1 Introduction

The identification and characterization of network traffic is one of the most important activities for an operator. Through the continuous monitoring of the traffic, security policies can be deployed and tuned, anomalies can be detected, changes in the users behavior can be identified so that QoS and traffic engineering policies can be continuously improved.

In the last years, several traffic classification techniques have been proposed. At the beginning *port-based* approaches were mainly used; however, the characteristics of many nowadays applications that employ randomly chosen ports, significantly reduce the effectiveness of these approaches [1–4]. Those are today abandoned in favor of *deep packet inspection* (DPI) or *behavioral* techniques [13, 15]. In the first case, the traffic is classified looking for specific keywords inside the packet payload, e.g., `BitTorrent` or `GET/POST` keywords identify the BitTorrent and HTTP protocols, respectively. Behavioral techniques try to overcome the limitations of DPI, e.g., when payload is encrypted, by exploiting some description of the application behavior through statistical characteristics, such as the length of the first packets of a flows.

All these classifiers share some key aspects. On the one hand a deep domain knowledge is required to correctly train and periodically update these classifiers. On the other hand, the classifiers can identify only the specific applications they have been trained for; all other traffic is aggregated in a single class labeled

J. Domingo-Pascual, Y. Shavitt, and S. Uhlig (Eds.): TMA 2011, LNCS 6613, pp. 150–163, 2011.
© Springer-Verlag Berlin Heidelberg 2011

as "unclassified". The classifiers are therefore typically tuned to identify the prominent classes but they completely miss the dynamics of the rest of the traffic. For example, they cannot identify the introduction of a new application, or changes in the users' behavior or in the applications protocols.

Classification can happen at different degrees of granularity: *packet*, *flow*, or *endpoint*[1], with significant differences on the number of objects to be considered. However, when mining the subset of unclassified traffic, the number of objects to be analyzed is still large even when considering higher aggregation levels. For instance, for moderate traffic aggregates, the even small fraction of unclassified traffic is typically built by thousands of endpoints, each aggregating tens of flows made of hundreds of packets. How to practically reduce the number of unknown objects to analyze is therefore a key problem.

In this paper, we focus our attention on the inspection of the unclassified traffic. We propose an unsupervised technique that, having no knowledge of the applications that generate the traffic, partitions a traffic aggregate into "clusters" that are distinguished based on common features, i.e., they exhibit a common treat. A simple clustering methodology based on the K-means algorithm is augmented with the capability to effectively determine the number of traffic clusters K. The results is a simple algorithm that can reduce the number of objects to analyze to few tens, even if the total traffic amounts to several tens of megabits per seconds. By being completely automatic and unsupervised, the proposed methodology can be engineered to: i) identify new classes of traffic by exploiting the network administrator domain knowledge when inspecting a traffic cluster; ii) monitor the traffic evolution by highlighting the birth of traffic clusters corresponding to traffic of previously unobserved applications; iii) design anomalies detection techniques by observing the evolution of traffic clusters over time.

To test and validate our methodology, we consider some UDP traffic traces of which we already have a deep knowledge on, achieved through a combination of DPI and statistical techniques, as well as the results of some active experiments. We consider UDP traffic since today its importance is steadily increasing [4], and few works explicitly targeted it in the past. We apply the proposed technique to the traces and check the coherence of the automatic classification with our ground truth. Experimental results show that the proposed clustering algorithm is very effective. Clusters accuracy is typically higher than 95% and the number of clusters is also very small, e.g., never larger than 40, and typically in the order of 25. Such a good performance is due to both the descriptiveness of the KISS features, and the goodness of the agglomerative process. With respect to previous proposals [5, 6] in which hundreds of clusters were needed to achieve good accuracy, the major advantage of our solution is that it reduces the time needed to inspect the clusters since the traffic is better partitioned.

Finally, we present some examples of classification of unknown traffic we were able to identify.

---

[1] A flow is commonly defined as the group of packets that have the same tuple {srcIP, dstIP, srcPort, dstPort, protocol}. An endpoint identifies the group of flows having the same {host IP, host Port, protocol} tuple.

2   Data Mining Techniques and Related Work

Machine learning algorithms are data mining techniques used to create a model from a dataset. They can be grouped in two families: *supervised* and *unsupervised* techniques. In both cases, objects are characterized by features, i.e., a vector of characteristics that can be extracted automatically by observation. Supervised algorithms exploit a training dataset in which each object is labeled, i.e., it is a-priori associated to a particular class. This coupling is used to create a suitable model so that objects with the same labels are grouped together. Then, unlabeled objects can be associated to a class previously defined according to their features. For unsupervised algorithms, instead, the grouping operation is automated without any knowledge of a-priori labels. Groups of objects are then clustered based only on a notion of distance evaluated among samples, so that objects with similar features are part of the same cluster. Supervised algorithms allow high accuracy during classification, provided that the training set is representative of the objects.

The application of machine learning techniques is not new in the traffic classification field. [7] is one of the preliminary works and shows that clustering techniques are useful to obtain insights about the traffic. In [6] supervised and unsupervised techniques are compared, demonstrating that unsupervised algorithms can achieve the same performance of the supervised algorithms. Other works compare the accuracy of different unsupervised algorithms [3, 5, 8]. In general, the techniques presented in these works achieve a very high accuracy but they typically require several hundreds of clusters, therefore making it difficult to then inspect and label the clusters. Recently, [9] and [10] have introduced the *semi-supervised* methodology. They exploit the advantages of both methodologies: a clustering algorithm is used to partition the dataset as in the unsupervised case. Part of the dataset is labeled, so that it is possible to extend the classification to all objects in the same cluster. Results shows that the accuracy of the classification largely depends on the goodness and coverage of the labeled dataset, and clusters without labeled objects cannot be further classified.

All previous works focus on the classification accuracy of some target classes, i.e., a small subset of the applications to consider. Real traffic is however composed by a large mix of applications and often it is crucial to mine the remaining part of the traffic which is still unclassified. For example, in [3] authors show that the best classifier has poor performance when considering the unclassified traffic which amounts to more than 10% of the total.

In addition, the Internet represents a dynamic environment in which new applications are born, evolve and die continuously. By following these patterns, it is possible to better understand the users behaviors and the technology trends.

3   Feature Selection: Kiss Signatures

Machine learning algorithms are based on a description of objects summarized in a vector. The elements of the vector are called *features* and constitute a description of all known characteristics of the instance. They play a key role in the effectiveness of the machine learning algorithm, i.e., the more descriptive the features are, the better the performance is. In the past, most of the works considered a large set of generic features, such as packet/flow length, port number, round trip time. In this paper, instead, we rely on the signatures defined by Kiss, a stochastic classifier that we proposed in [11, 12]. The intuition behind Kiss is that application-layer protocols can be identified by statistically characterizing the stream of bytes observed in a flow of packets. Kiss builds protocol signatures by measuring the randomness of groups of bits extracted from the packets payload. Considering an analogy, this process is like recognizing the foreign language by considering only the cacophony of the conversation, i.e., by letting the protocol format emerge, while discarding its actual semantic. Kiss features proved to be highly descriptive when adopted in supervised machine learning algorithms for traffic classification.

Kiss signatures are computed over the packets directed to or originated from a given *endpoint*. They aim at measuring the randomness of the first bytes of the packet payload that are those usually carrying application header. In particular, the first 12 bytes of the packet payload are divided into groups of $b = 4$ bits, for a total of $G = 24$ groups. For each group, the statistic of the occurrence of each of the $2^b = 16$ possible values is computed over $N = 80$ packets. Then, the randomness of each group g, denoted by X_g, is measured as the Chi-Square distance of the group statistics with respect to the uniform distribution,

$$X_g = \sum_{i=0}^{2^b-1} \frac{(O_i^g - E_i)^2}{E_i} \tag{1}$$

where O_i^g is the observed occurrence of the value i for the g group, and $E_i = N/2^b$ is the expected occurrence for the uniform distribution. Finally, since the value of X_g grows exponentially with the number of deterministic bits in the group, and linearly with N [11], we derive,

$$b_g = \log_2 \left(\frac{X_g}{N} + 1 \right) \tag{2}$$

where b_g represents then the number of constant bits in group g. The vector $\{b_1, b_2, \ldots, b_G\}$ represents the Kiss signature used in the rest of the paper.

Since Kiss features are obtained from inspection of packet payload, encrypted application layer protocols may limit the goodness of the features, i.e., all groups may look like random data. In those cases, it would results impossible to correctly distinguish two applications that adopt fully encrypted payload.

In summary, each Kiss signature computed from the traffic of an endpoint corresponds to a "point" in an hyper-space of 24 dimensions. Given then a set of monitored endpoints and of corresponding Kiss signatures, the objective of this work is to identify "clouds" of similar points, i.e., to clusterize the signatures with no a-priori knowledge about the applications that generated the traffic. Resulting clusters are higher level objects that can be investigated further, and whose properties naturally extend to each endpoint in it. For example, the clusters give indications about how the traffic volume distributes among the regions suggesting which dataset should be further inspected. Similarly, by constantly monitoring the galaxy of clouds in time, it would also be possible to identify traffic shifts due to the rise/decline of applications, to the presence of anomalous behaviors, or malicious users.

4   Clustering Methodology

Kiss signatures map the traffic generated by applications into points in an hyper-space. To partition the space into pure clusters where points are generated by the same application, we leverage on the *K-means* algorithm, a classic unsupervised technique [13]. Given a set of K "centroids", the K-means algorithm iterates over two steps: it first assigns each point to the closest centroid, defining a cluster; then, each cluster centroid is re-computed as the arithmetic mean among all points of the cluster. The algorithm ends either after a predefined number of iterations or if centroids do not change at a given iteration. At the beginning, centroids are randomly picked.

The major drawback of K-means is that it assumes the a-priori knowledge of the number K of clusters one is interested in. The proposed algorithm tries to overcome this limitation using an agglomerative approach. We start by decomposing the hyper-space in a large number of clusters, K_0. Then, we incrementally merge the two closest clusters until one cluster only remains. A similar technique was successfully applied to the network measurement context in [14]. The pseudo-code of the algorithm is:

```
K = K0
centroids, labels = K-means(K, data)
while (K > 1)
    c1, c2 = closest_centroids(centroids)
    centroids = merge_centroids(centroids, c1, c2)
    labels = redo_labeling(data, centroids)
    K = K - 1
```

We start running K-means with $K_0 = 100$ randomly chosen centroids, i.e., we force the partitioning of the hyper-space in a large number of small clusters that are extremely pure. The algorithm then iterates merging at each step the two closest clusters: at step K, the algorithm looks for the two closest centroids $c1, c2$, it merges them into a new centroid positioned at the geometric barycenter of $c1$ and $c2$; then points are reassigned to the new set of $K - 1$ centroids. The algorithm continue the aggregation until 2 clusters only remain.

The rationale behind the algorithm is that two centroids which are very close are likely to be associated to the same final cluster. By monitoring the value of the closest distance between centroids at each iteration step, and using this as an *indicator function*, it is possible to decide the optimal value of K, namely K_c, to stop at.

In our scenario, K_c represents the estimated number of protocols that are present in the dataset. Let the smallest distance between centroids be defined as

$$\gamma_K = (d_K - d_{K-1})^2 \tag{3}$$

where d_K is the Euclidean distance between the two closest centroids at step K of the algorithm. γ_K defines our indicator function. Since the distance between points (and clusters) that correspond to the same protocol is expected to be smaller than the distance between points that correspond to traffic generated by different applications, large values of γ_K suggest that the algorithm is artificially enforcing the merging of two clusters that are quite different from each other.

Notice that only a single run of K-means is executed at the beginning to obtain the initial set of clusters. At each iteration, the algorithm works only on the centroids, and this has two main benefits. First, we can better control the modification on the space due to aggregation. In fact, the K-means algorithm is subject to "oscillation effect", i.e., small modifications in the centroid position could lead to large transformations in the cluster geometry. By using centroids only we avoid to re-assign samples to centroids, so that the quality of the initial clustering is better preserved.   In addition, by considering centroids only we reduce the computational cost by several order of magnitudes, we handle $O(K_0)$ centroids instead of $O(N)$ samples ($N >> K_0$). Moreover, the K-means complexity depends on the maximum number of iterations I (which in our case we set to 100), so that its complexity is $O(IN)$. In our experiments on an AMD Athlon-64 X2 Dual Core Processor 4200+, we elaborated several thousands of points present in a 15 minute long traffic traces in less than 3 minutes, the largest majority of the time being devoted to the initial K-Means run. Given that the code used can be further optimized, the result is promising and suggests that the algorithm might be applied to real-time monitoring.

Finally, K-means is known to suffer from the choice of the initial centroid position. Usually initial centroids are randomly chosen so that different starting conditions can lead to different clustering. In our scenario, since we select a large number K_0 of clusters, the bias introduced by the selection of the initial centroids is minimal. We performed some tests by running the algorithm with different initial random seeds and the results (not reported for the lack of space) show that there is practically no influence on the initial choice.

5   Experimental Results

5.1   Datasets

The results presented in this paper refer to datasets extracted from two traces, called *ISP-Trace* and *P2PTV-Trace*; the traces are described in Table 1.

ISP-Trace is a real traffic trace collected from the network of an Italian large ISP which offers converged services, in which data, native VoIP, and IPTV share a single broadband connection. This dataset is representative of a very heterogeneous scenario, in which users are free to use the network without any restriction.

Table 1. Description of the ISP-Trace (a) and P2PTV-Trace (b)

	Protocol	#flows ×10³ (%)		Mbytes (%)		#endp. ×10³ (%)		#sign. ×10³(%)	
(a)	BitTorrent	217	(3.39)	40	(0.19)	34	(4.14)	22	(0.33)
	DNS	260	(4.05)	185	(0.88)	153	(18.79)	31	(0.47)
	eMule	5200	(80.96)	936	(4.43)	476	(58.56)	61	(0.91)
	RTCP	8	(0.13)	46	(0.22)	6	(0.73)	25	(0.38)
	RTP	9	(0.14)	18244	(86.26)	7	(0.86)	6222	(92.14)
	Unclassified	728	(11.34)	1698	(8.03)	137	(16.92)	390	(5.78)
	tot	6422	(100.00)	21149	(100.00)	813	(100.00)	6751	(100.00)
(b)	PPLive	27	(78.52)	1585	(32.96)	184	(38.90)	23	(28.30)
	SopCast	5	(14.87)	2282	(47.43)	176	(37.21)	48	(57.46)
	TVants	2	(6.61)	943	(19.61)	113	(23.89)	12	(14.24)
	tot	34359	(100.00)	4810	(100.00)	473	(100.00)	83	(100.00)

It therefore is a very challenging scenario for traffic classification. In this paper we present results considering a dataset obtained monitoring a PoP for 24 hours in October 2007, during which about 21GB of UDP traffic and 813,000 endpoints were monitored. Some *known* protocols (BitTorrent, eMule, RTP, RTCP and DNS) have been extracted from the aggregated trace using Tstat [15], a traffic classifier that combines a number of DPI mechanisms with statistical techniques. The classification has been manually cross-checked to have a high confidence in the ground truth. These protocols account for more than 90% of the total volume, as shown in Table 1. The remaining 10% of traffic has been labeled as "unclassified".

P2PTV-Trace was collected during ad-hoc experiments that were organized to observe the performance of popular P2P-TV applications, namely PPLive, SopCast and TVants. The resulting dataset [16] consists of packet level traces collected from more than 45 PCs running P2P-TV applications in 5 different Countries, and it is representative of a wide range of different scenarios. Being the result of active experiments, the trace contains only a single protocol at a time and we have a perfect knowledge about it.

The datasets extracted from the two traces are disjoint. In fact, there is no P2P-TV traffic in the ISP-Trace. When needed, we can artificially "inject" P2P-TV traffic from the P2PTV-Trace into the ISP-Trace to increase the number of known protocols when assessing the performance of the clustering algorithm.

5.2   Evaluation of the Proposed Approach

Fig. 1(a) shows the evolution of the indicator function during the application of the algorithm to ISP-Trace considering a 10 minute long trace. The minimum distance between clusters is very small for values of $K > 20$, suggesting that the algorithm is merging clusters whose centroids are very close. Instead, for $K \leq 20$, the algorithm starts merging cluster centroids which are quite far from each other, suggesting an improper and artificial merging.

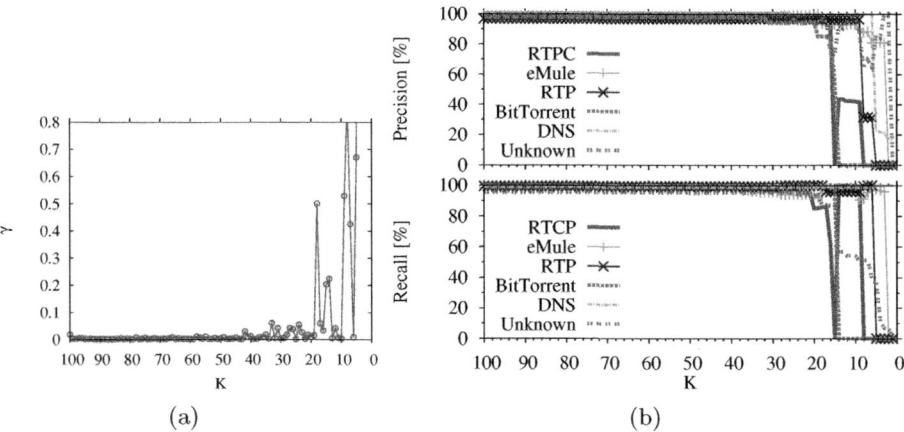

Fig. 1. Evolution of the clustering algorithm: the indicator function (a) and classification accuracy in terms of precision and recall (b)

To confirm this intuition, the homogeneity of each cluster is evaluated against the endpoint classification obtained by Tstat (our ground truth). Fig. 1(b) reports the precision (top) and recall (bottom) performance indexes, defined as

$$Precision = \frac{true\ pos}{true\ pos + false\ pos} \qquad Recall = \frac{true\ pos}{true\ pos + false\ neg} \qquad (4)$$

for different values of K. Precision is a measure of exactness or fidelity, whereas recall is a measure of completeness; these two measures complement each other. A precision of 1.0 for a class C means that every item labeled as belonging to C does indeed belong to C. It however says nothing about the number of items from class C that were not labeled correctly. A recall of 1.0 means that every item from class C was labeled as belonging to class C. It however says nothing about how many other items were incorrectly labeled as belonging to class C.

Consider Fig. 1(b); two observations hold. First, for $K > 20$, the fidelity and completeness of the identified clusters is very high, proving that the Kiss signatures accurately represent different protocols, and that traffic generated by different applications can be easily clustered. Second, the abrupt decrease of both precision and recall observed in Fig. 1(b) for $K \leq 20$ confirms that some clusters corresponding to different protocols are artificially merged, causing the formation of impure clusters.

To further assess the goodness of the approach, we inspect the behavior of the indicator function considering datasets in which we progressively add traffic of various applications. We start by considering a dataset containing only Sop-Cast and TVants traffic; we then add, in sequence, the traffic of PPLive, RTP, BitTorrent, DNS and eMule to the dataset. For each traffic mix we run our algorithm. The results are reported in Fig. 2 for $K \leq 20$, only. The figure shows that the indicator function abruptly increases for values of K that are strongly related with the number of traffic classes. A simple thresholding mechanism on

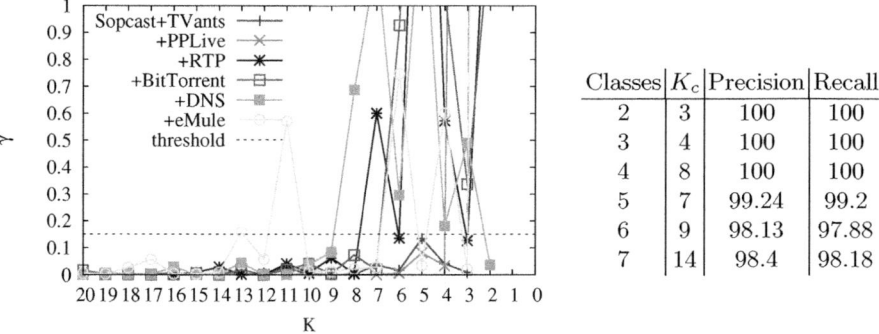

Fig. 2. Example of indicator function for traffic aggregates with progressively increasing number of classes

Classes	K_c	Precision	Recall
2	3	100	100
3	4	100	100
4	8	100	100
5	7	99.24	99.2
6	9	98.13	97.88
7	14	98.4	98.18

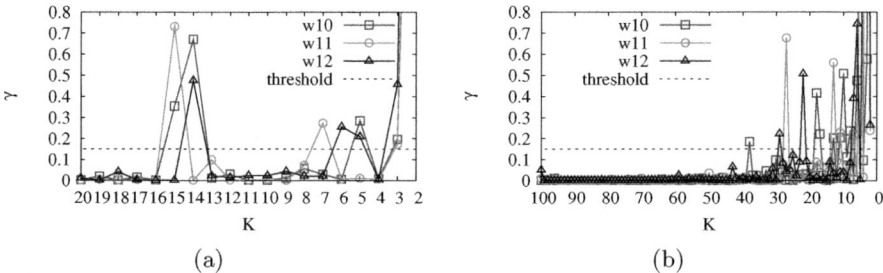

(a)   (b)

Fig. 3. Evolution of the clustering algorithm over different time windows of the ISP-Trace without (a) and with (b) the unclassified traffic

the indicator function can be adopted to automatically detect the value K_c. As an example, the figure reports a threshold of 0.15 that resulted very effective in our tests.

The Table on the right of Fig. 2 reports the suggested number of clusters K_c obtained with the threshold $\gamma = 0.15$, the corresponding recall and precision are also indicated. Results confirm that the value of K_c increases with the number of traffic classes. The resulting precision and recall are extremely high, and a marginal decrease is observed only when considering more than 5 protocols. This is due to BitTorrent traffic which is sometimes confused with TVants traffic whose Kiss signatures result similar. Nevertheless, the performance are very good.

Interestingly, the number of identified clusters is larger than the actual number of applications. This is due to single applications using multiple protocols with different formats, e.g., signaling is different respect to data messages. The Kiss signatures are therefore different, and the clustering algorithm correctly identifies separate clusters.

Finally, we repeat the experiment considering other different 10-min-long traces extracted from the ISP-Trace. The goal is to investigate if the indicator function always correctly suggests the number of cluster to use. Fig. 3(a) reports the indication functions obtained for the three windows considering the aggregation of the 7 considered traffic classes. No unclassified traffic is present. We can see that the suggested K_c is consistent among all the experiments. Instead, Fig. 3(b) reports the indicator functions when unclassified traffic is present too. In this case, since different traffic mixes are present during different periods of time, higher noise is present with respect to the previous case and different values K_c are selected in different windows. In conclusion, the indicator function suggests an optimal number of clusters which changes depending on the actual traffic mix. Thus, a conservative large number of clusters is preferable, especially when considering different time windows. Moreover, note that in Fig. 3(b) the number of suggested clusters is never higher than 40.

5.3   Comparison with Other Clustering Techniques

The automated selection of the optimal number of clusters is not new in literature. Several score indexes have been proposed to precisely correlate the goodness of the clustering with the number of used clusters. Examples of these indexes are: the Bayesian Information Criterion (BIC) adopted by the XMeans algorithm [13] and the Normalized Mutual Information (NMI) [3]. In this work, we are interested in investigating the automated approaches which do not require the a-priori knowledge of the points' labels (that is instead required by the NMI). We evaluated the performance of both XMeans and NMI; in addition, we considered also the DBScan algorithm. XMeans shows similar performance as our algorithm in terms of recall and accuracy. However, the number of identified clusters is typically much larger than the one obtained by our algorithm. For example, XMeans accuracy is higher than 95%, but at least 10 more clusters are identified, i.e., 50% more than with our proposal. Considering NMI, the accuracy is lower than 95% when 25 clusters are used, as suggested by the NMI technique. With 40 clusters, performance of the NMI-based method is similar to the one of our algorithm. Finally, DBScan performed poorly achieving only 85% of accuracy with the best parameter setting.

Notice also that all previous algorithms are computationally more expensive than our proposal. In conclusion, the proposed algorithm is completely automated, does not require any knowledge of the points labels and seems a good trade-off among clustering accuracy, number of clusters and complexity.

5.4   Clusters Distances

In this section, we investigate the geometry of the clusters of points identified by our algorithm. The results presented in this section are obtained using $K = 40$ (a conservative large value) and refer to a single time window of ISP-Trace. For the other time windows, not shown here for the sake of brevity, we obtained similar results.

We start by considering the size of each cluster. Fig. 4 shows the cumulative distribution function of the normalized Euclidean distance between each point and its centroid. As we can see, half of the points in the dataset are very close to

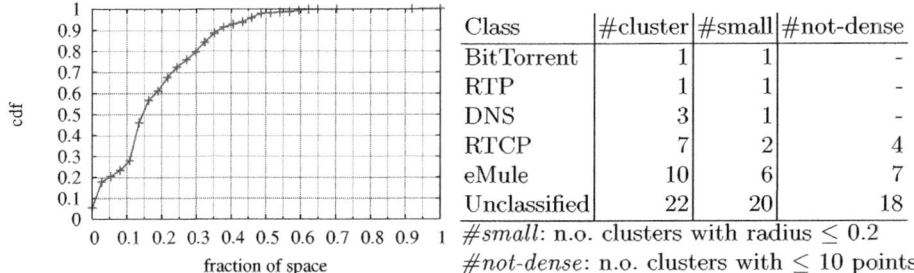

Class	#cluster	#small	#not-dense
BitTorrent	1	1	-
RTP	1	1	-
DNS	3	1	-
RTCP	7	2	4
eMule	10	6	7
Unclassified	22	20	18

#*small*: n.o. clusters with radius ≤ 0.2
#*not-dense*: n.o. clusters with ≤ 10 points

Fig. 4. Cumulative distribution function of the distance between points and centroid in each cluster (plot on the left) and a few additional data (table on the right), obtained running the algorithm on ISP-Trace with $K = 40$

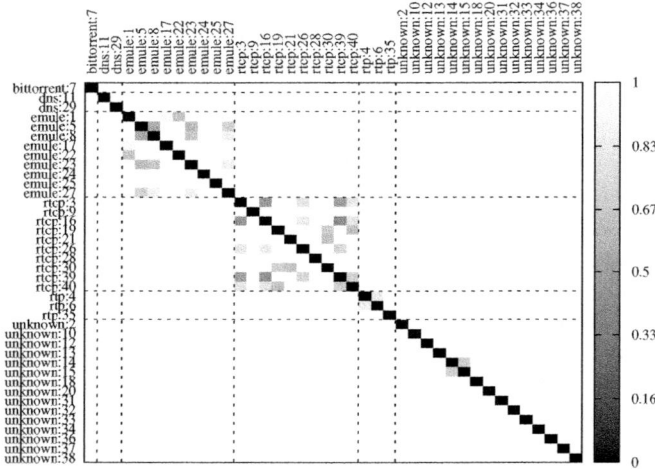

Fig. 5. Distance between centroids of different clusters, for 40 clusters obtained by running the algorithm on IPS-Trace with $K = 40$

theirs centroid, with a distance smaller than 20% of the cluster space size. The table on the right of Fig. 4 reports some statistics about the clusters geometry according to the DPI classification. In particular, the second column reports the number of clusters identified for each application, the third column reports the number of *small* clusters, i.e., clusters with a radius smaller that 0.2, and the last column gives the number of *not-dense* clusters, i.e., clusters with less than 10 samples. BitTorrent and RTP are mapped into a single cluster, while the "un-classified" is composed of a set of small, often not-dense, clusters. Interestingly, eMule is highly partitioned too. Investigating further, we noticed that each cluster corresponds to a different protocol which eMule uses for different purposes, e.g., one protocol is used to exchange messages with the server, another one is used to exchange traffic with peers.

To better understand the possible overlapping between the clusters, Fig. 5 reports the distance between pairs of centroids. The distance has been mapped

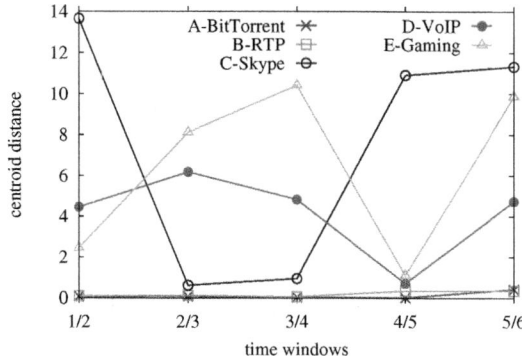

Fig. 6. Example of evolution of the centroids position

onto a gray scale map in which the darker the color is, the nearer the centroids are. The image is symmetrical with respect to the main diagonal, where all points have a distance of 0 by definition. Clusters are ordered based on their type of traffic so that clusters referring to the same application are nearby; dashed lines are used to delimit the applications. The only blocks which include nearby clusters are related to the same application.

In conclusion, we can say that the Kiss signatures map different protocols in different compact clusters of the hyper-space. The geometry of the clouds is strictly related to the characteristics of the application, but the signatures are naturally clustered in pure areas which do not overlap.

6 Mining the Unclassified Traffic

In this section we show how the proposed technique can be used to monitor the traffic evolution in time and detect the presence or absence of traffic in different periods. To do so, we measure the modifications of the clouds obtained by running the algorithm over consecutive time windows. We consider 1 hours of traffic divided into six 10-min-long traces and for each trace we run our algorithm using $K = 40$, as previously described. The centroids obtained for each time window are then compared with centroids identified in the previous time window. Each centroid is associated to the closest cluster in the previous set according to their geometric distance. This allows to detect changes between the current and the previous cluster placement.

Fig. 6 reports some interesting examples; it shows the distance of some selected centroids in consecutive time windows. For example, the position of centroid A and centroid B is practically the same over time. Verifying the corresponding clusters, we found out that samples of cluster A and cluster B are associated to BitTorrent and RTP, respectively; since in the traffic traces those applications are always present, the corresponding centroids are always present and more or less in the same position.

Consider now the case of the cluster with centroid C. The minimum distance among the centroid C in the first and the second time window is very high, suggesting that in the second time window C is associated to a cluster of traffic that was not present during the previous time window. When comparing centroid C to its closest centroid at time window 3, we see that it moved very little. Similarly, considering time windows 3 and 4, centroid C is still referring to the same cloud of samples. Only in time window 5, the centroid C seems to disappear, since the closest centroid is very far from its position during time window 4. This suggests that some new traffic appears at the 2-nd time window, it is present during the 3-rd and 4-th time window, when it disappears again. Investigating further, we discovered that the traffic was generated by a Skype call that lasted for that period of traffic. Centroid C then refers to Skype Voice protocol.

Similar conclusions can be drawn following centroid D and centroid E evolution. Comparing their position during the 4-th and the 5-th window, we can observe that they moved little, i.e., they refer to the same cluster. Manual inspection revealed that the traffic of cluster-D corresponds to STUN protocol - Simple Traversal of User Datagram Protocol that was initiated by some P2P client that was alive in time window 4 and 5. Centroid E refers, instead, to traffic between hosts that used port 16567. This latter is composed by both short packets and much bigger packets, which might be related to Battlefield2 protocol.

Beside these examples, the methodology identified other sets of clusters and centroids which were always placed in the same zone across consecutive the windows. Some of these clusters were due to long-lived, single connections carrying many bytes, while others contained P2P-like flows, i.e. endpoints exchanging limited amount of data with an large number of hosts. Unfortunately, because of the limited amount of available payload, we are not able to further identify the application that generated these flows.

These examples show how we could successfully employ our technique to get insights into the unclassified traffic that Tstat DPI and behavioral classifiers cannot identify. In terms of traffic volumes, we could correctly identify and clusterize more than the 40% of unclassified traffic.

7   Conclusion

In this paper, we presented a clustering methodology to partition a traffic aggregate in classes according to the generating application. Using statistical signatures as those of Kiss, one of our classifiers, the methodology (that is completely unsupervised) is based on the K-Mean clustering algorithm enhanced through a mechanism to detect the optimal number of clusters.

Results show that the traffic partitions are very accurate. and confirm that the statistical signatures are effective in capturing the differences among application protocols. Moreover, our results prove that the methodology can be effectively used in different contexts. First of all, it is helpful to mine the unclassified traffic, i.e. the traffic that traditional DPI or a behavioral classifiers cannot recognize. Indeed, it helped us revealing 40% of the traffic we could not classify with our classifiers. Second, the algorithm can reveal the born of new applications, as well as the changes of existing ones.

References

1. Karagiannis, T., Broido, A., Brownlee, N., Claffy, K., Faloutsos, M.: Is p2p dying or just hiding? In: Globecom, Dallas, TX (November 2004)
2. Karagiannis, T., Papagiannaki, D., Faloutsos, M.: Blinc: Multilevel traffic classification in the dark. In: SIGCOMM, Philadelphia, PA (August 2005)
3. Bernaille, L., Teixeira, R., Salamatian, K.: Early application identification. In: CoNEXT, Lisboa, PT (December 2006)
4. Zhang, M., Dusi, M., John, W., Chen, C.: Analysis of UDP Traffic Usage on Internet Backbone Links. In: Proceedings of the 2009 Ninth Annual International Symposium on Applications and the Internet, Seattle, WA (July 2009)
5. Erman, J., Arlitt, M., Mahanti, A.: Traffic classification using clustering algorithms. In: ACM SIGCOMM, Pisa, IT (September 2006)
6. Erman, J., Mahanti, A., Arlitt, M.: Internet traffic identification using machine learning. In: IEEE GLOBECOM, San Francisco, CA (December 2006)
7. McGregor, A., Hall, M., Lorier, P., Brunskill, J.: Flow clustering using machine learning techniques. In: Barakat, C., Pratt, I. (eds.) PAM 2004. LNCS, vol. 3015, pp. 205–214. Springer, Heidelberg (2004)
8. Wang, Y., Xiang, Y., Yu, S.: An automatic application signature construction system for unknown traffic. Concurrency and Computation: Practice and Experience 22, 1927–1944 (2010)
9. Erman, E., Mahanti, A., Arlitt, M., Cohen, I., Williamson, C.: Semi-supervised network traffic classification. In: ACM SIGMETRICS, San Diego, CA (June 2007)
10. Yuan, J., Li, Z., Yuan, R.: Information entropy based clustering method for unsupervised internet traffic classification. In: IEEE ICC, Beijing, CN (May 2008)
11. Finamore, A., Mellia, M., Meo, M., Rossi, D.: KISS: Stochastic Packet Inspection Classifier for UDP Traffic. IEEE/ACM Transactions on Networking 18(5), 1505–1515 (2010)
12. Mantia, G.L., Rossi, D., Finamore, A., Mellia, M., Meo, M.: Stochastic Packet Inspection for TCP Traffic. In: IEEE International Conference on Communication - ICC, Cape Town, SA (May 2010)
13. Berkhin, P.: A survey of clustering data mining techniques. In: Kogan, J., Nicholas, C., Teboulle, M. (eds.) Grouping Multidimensional Data, ch. 2, pp. 25–71. Springer, Heidelberg (2006)
14. Bianco, A., Mardente, G., Mellia, M., Munafò, M., Muscariello, L.: Web User-session Inference by Means of Clustering Techniques. IEEE/ACM Transactions on Networking 17(2), 405–416 (2009)
15. Finamore, A., Mellia, M., Meo, M., Munafò, M., Rossi, D.: Live Traffic Monitoring with Tstat: Capabilities and Experiences. In: Osipov, E., Kassler, A., Bohnert, T.M., Masip-Bruin, X. (eds.) WWIC 2010. LNCS, vol. 6074, pp. 290–301. Springer, Heidelberg (2010)
16. Ciullo, D., da Rocha Neta, A.G., Horvath, A., Leonardi, E., Mellia, M., Rossi, D., Telek, M., Veglia, P.: Network Awareness of P2P Live Streaming Applications: a Measurement Study. IEEE Transanctions on Multimedia 12(1), 54–63 (2010)

Entropy Estimation for Real-Time Encrypted Traffic Identification (Short Paper)

Peter Dorfinger[1], Georg Panholzer[1], and Wolfgang John[2]

[1] Salzburg Research, Salzburg, Austria
{peter.dorfinger,georg.dorfinger}@salzburgresearch.at
[2] Chalmers University of Technology, Göteborg, Sweden
wolfgang.john@chalmers.se

Abstract. This paper describes a novel approach to classify network traffic into encrypted and unencrypted traffic. The classifier is able to operate in real-time as only the first packet of each flow is processed. The main metric used for classification is an estimation of the entropy of the first packet payload. The approach is evaluated based on encrypted ground truth traces and on real network traces. Encrypted traffic such as Skype, or encrypted eDonkey traffic are detected as encrypted with probability higher than 94%. Unencrypted protocols such as SMTP, HTTP, POP3 or FTP are detected as unencrypted with probability higher than 99.9%. The presented approach, named real-time encrypted traffic detector (RT-ETD), is well suited to operate as pre-filter for advanced classification approaches to enable their applicability on increased bandwidth.

Keywords: entropy estimation, real-time detection, traffic filtering.

1 Introduction

During the last years the diversity of web based applications and their traffic patterns has increased enormously. This hinders network management activities which are to a substantial extent based on discriminative treatment of application traffic. Available systems for traffic classification are either based on fast but vague matching of IP addresses, transport protocol and ports or complex *deep packet inspection* (DPI) or statistical approaches. Especially applications that hide the traffic within encrypted communication are difficult to detect.

To detect such hidden applications often complex algorithms have to be performed. Due to the complexity such approaches have to inspect all packets on a link and consequently are unable to operate on high bandwidth links.

To overcome these shortcomings a fast and simple approach is needed that is able to pre-filter the traffic. The pre-filtering has the advantage to reduce the traffic which has to be further inspected to a reasonable amount. To keep the complexity of the approach low, it is a prerequisite to minimize status information. Thus our approach is designed such that pre-filtering of the traffic is performed upon arrival of the first communication packet (excluding TCP 3way handshake).

J. Domingo-Pascual, Y. Shavitt, and S. Uhlig (Eds.): TMA 2011, LNCS 6613, pp. 164–171, 2011.
© Springer-Verlag Berlin Heidelberg 2011

The classification based on the first packet solely has, beside the aspect of keeping status information low, the main advantage to be operational in real-time.

2   Related Work

In the late 1990s traffic classification was mainly performed on well known port numbers. Nowadays traffic classification is more and more getting complicated as an increasing number of applications try to hide its traffic from detection or classification. As hiding is often performed within encrypted network traffic, entropy-based classification algorithms have gained interest within the last years. Entropy-based approaches are often used to detect malicious traffic [1,2].

Lyda and Hamrock [1] use an entropy-based approach called bintropy that is able to quickly identify encrypted or packed malware. The entropy is used to identify statistical variation of bytes seen on the network.

Pescape [3] uses entropy to reduce the amount of data that has to be processed by traffic classification tools. Entropy is used as input for an advanced sampling approach that ensures that sensible information needed to get an appropriate model of the network traffic is still present. Packets not needed for an appropriate model are dropped.

An entropy-based approach which inspired the present work is presented in [2]. The N-truncated entropy for different encrypted protocols is determined. For example, for an HTTPS connection the byte entropy after 256 bytes of payload should be between six and seven. If the value for a specific connection is below this range, it is assumed that the connection is subverted.

In an earlier work [4] we concentrated on the possibility to pre-filter Skype traffic based on information gathered from the first packet of a flow.

3   Entropy and Entropy Estimation

In 1948 Shannon [5] developed a measure for the uncertainty of a message. This measure is known as entropy in information theory. Shannon considers the case where we have a fixed number m of possible events $A_1, \ldots, A_m$ whose probabilities of occurrence $p_1, \ldots, p_m$ are known.

Entropy is defined as

$$H = -K \sum_{i=1}^{m} p_i \log(p_i), \qquad (1)$$

Entropy can be used as a measure of the information content of a message. Equal probabilities lead to a maximum value for the entropy. Two concepts within data processing result in a high value of entropy. First data compression, as the bits needed for data representation should be minimized. Second data encryption, as any predictable behaviour available in the source data has to be removed. Both processing steps end with a data stream with equal probabilities for each event/symbol.

Within this work we are using entropy as a measure of uniformity. For an entropy-based test for uniformity an appropriate estimator for the entropy is needed. But as stated in literature [6] estimating the entropy based on a sample is hard to retrieve, especially for $N < m$. Consequently we focus on a uniformity test which is influenced by entropy but does not need the estimation of the actual entropy.

Olivain and Goubault-Larrecq [2] present a work where, motivated by the problems of estimating the entropy based on a sample of length N accordingly, the N-truncated entropy is used instead. The N-truncated entropy $H_N(p)$ is defined as follows. Generate all possible words w of length N according to p. Estimate the entropy based on maximum likelihood (MLE) for all words w. The N-truncated entropy is then the average of the MLE estimates, i.e. sum up all MLE estimates and divide by the number of words to retrieve $H_N(p)$.

According to [2] $H_N(p)$, given that $p_i = 1/m$ for all i, which means that p follows the uniform distribution $\mathcal{U}$, $H_N(\mathcal{U})$ can be calculated by

$$H_N(\mathcal{U}) = \frac{1}{m^N} \sum_{n_1+\ldots+n_m=N} \left[\binom{N}{n_1 + \ldots + n_m} \times \left(\sum_{i=1}^{m} -\frac{n_i}{N} \log \frac{n_i}{N} \right) \right]. \quad (2)$$

The maximum likelihood estimator can be used as an unbiased estimator of H_N. Checking for uniformity is then straightforward. Based on a sample of length N estimate the entropy using MLE and compare the result to $H_N(\mathcal{U})$. The closer the estimated value is to $H_N(\mathcal{U})$ the more likely is it that the underlying distribution is uniform. Within our work we are using a Monte-Carlo method for estimating $H_N(\mathcal{U})$ and the confidence intervals.

For very short words this method for detecting uniformity fails. Paninski [7] states that for a uniformity test $N > \sqrt{m}$ samples are needed.

4   Classification

The proposed traffic classifier is based on a 2-stage approach, where the false-positive rate of the entropy estimation based classifier is reduced by the coding based classifier.

4.1   Entropy Estimation Based Classification

The entropy estimation based classification is the core classification component of the whole approach. In contrast to other approaches like [2], only the first packet that transports payload is used to identify encrypted flows. While on the one hand this makes it more difficult to calculate an accurate estimate of the entropy it enables the utilization of this technique in an online fashion, i.e., to identify encrypted flows in live traffic without the need to buffer or delay packets.

The basic concept of our approach is to estimate the N-truncated entropy of the actual payload $\hat{H}_{MLE}(w)$ and compare the result to the estimated entropy of uniformly distributed random payload $H_N(\mathcal{U})$ of the same length. The difference between these two estimates is used to decide whether the flow is identified as being encrypted or not. In accordance to Paninski [8] we do not use the entropy estimation for payload with less than 16 bytes.

In order to preserve most of the encrypted flows we decided to use $\hat{H}_N(\mathcal{U}) \pm 3 \times \sigma_{\hat{H}_N(\mathcal{U})}$ as a suitable confidence interval. This should include $\hat{H}_{MLE}$ of approximately 99.7% of uniformly distributed random payload.

4.2   Coding Based Classification

We assume that for plain text messages the payload is encoded in ASCII or ANSI where values from 32 to 127 are used for printable characters. Based on the entropy-based approach text message may look random, consequently we defined an algorithm that identifies large text blocks within the payload.

The probability that a character from a random source will be in the range from 32 to 127 is about 37.5%. Especially if at the beginning of the packet a large fraction of characters is in this range the payload is most likely unencrypted. Consequently we added a check that if the fraction of bytes with values between 32 and 127 is greater than 75% we assume that the flow is unencrypted.

As we want to reduce the processing time we do not evaluate the full payload of the packet. For the coding based classifier only the first 96 bytes are evaluated.

5   Algorithm

This section presents the usage of the entropy estimation and coding based classifiers within a novel approach for traffic classification based on information solely gathered from content of the first packet of a flow. The first packet in this context is defined as the first packet sent by a UDP connection and for TCP the first packet is the one that follows the 3 way handshake. The term flow is defined as bi-directional flow based on the 5-tuple.

We use a few lists to store information. The SYNList stores all 5-tuples where we have received a packet with the SYN flag set. A 5-tuple is removed from the SYNList after receiving the first packet containing payload or receiving a packet where the FIN flag is set. Furthermore the 5-tuple will be removed from the SYNList if, 60s after the SYN packet, there has not been any data packet. This should prevent the SYNList from growing due to SYN flooding attacks.

A TCP or UDP flow which was detected as encrypted will be stored in the ENCRList. The 5-tuple will be removed upon receiving a packet where the FIN flag is set. Unencrypted UDP flows are stored in the UNENCRList. The UDP 5-tuples are removed from the lists after 300s inactivity of this 5-tuple.

The first block in Figure 1 ensures that the TCP 3 way handshake for all flows will be forwarded, as traffic classifiers often need the 3 way handshake to identify the start of a TCP connection. The 5-tuple of this flow is stored in the SYNList. This behaviour can be changed to drop the 3 way handshake. The following two blocks are responsible for forwarding/dropping of packets belonging to flows that have already been identified as encrypted or unencrypted respectively. If a UDP

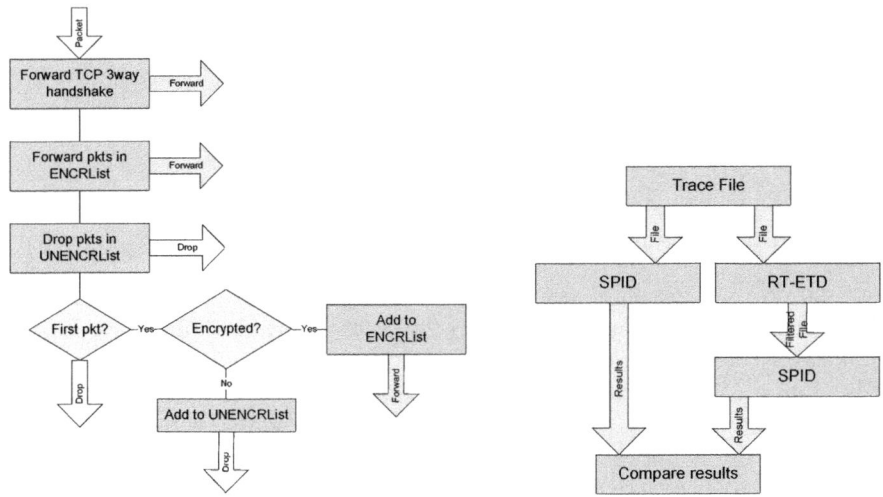

Fig. 1. Filtering flow chart   **Fig. 2.** Evaluation process

packet is not present in the ENCRList or in the UNENCRList it is the first packet of a flow. For TCP flows it is checked whether the 5-tuple of the packet is present in the SYNList, if so the packet is the first packet of a flow and has to be evaluated. The encrypted check executes entropy and coding based classifier to determine whether the packet/flow is encrypted or not. If the flow is encrypted it is added to ENCRList and forwarded, otherwise to the UNENCRList and dropped. Detailed information about the implementation can be found in [9].

6   Evaluation

For the evaluation we used another traffic classification tool (SPID, Statistical Protocol IDentification), together with real network traces and traces where the ground truth is known.

SPID [10] utilizes statistical packet and flow attributes to identify application layer protocols by comparing probability vectors of these attributes to known protocol models obtained on controlled training data. As comparison measure, the Kullback–Leibler divergence together with a threshold is used. SPID is a hybrid technique, utilizing generic attributes, which include statistical flow features (e.g. flow and packet lengths) as well as packet payload characteristics (e.g. byte frequencies and offsets). With a balanced combination of attributes, SPID was shown to be very effective in differentiating between encrypted and obfuscated protocols considered hard to classify [10].

As ground truth traces with encrypted traffic we use a subset of the fully classified traces of encrypted traffic also used within SPID [10]. Session of encrypted eDonkey, MSE and Skype protocols (Table 1) have been collected at a domestic

Table 1. Evaluation based on ground truth trace, note that more than 94% of the encrypted traffic is still present in the filtered file

Protocol	Flows	
	original	filtered
eDonkeyTCP encr.	398	94.5%
eDonkeyUDP encr.	828	99.6%
MSE	649	99.2%
SkypeTCP	91	97.8%
SkypeUDP	1973	98.0%

Table 2. Traffic shares of filtered file. The major fraction of the traffic belongs to encrypted protocols. SPID encr. includes all other encrypted protocols classifiable by SPID.

Traffic	amount
SYN + SYN/ACK flows	14.88%
VPN Key exchange	2.91%
VPN data	55.04%
Skype encr.	4.21%
SPID encr.	0.80%
AKAMAI	19.08%
unknown	3.08%

network connection in Sweden by using Proxocket[1], which enabled efficient separation of network traffic on a per-application basis.

Real network traces have been collected in a network of a small cable network provider in a segment used by about 100 customers. Several traces have been collected in this network, for the evaluation we are using three of them. A 1h/2GB trace, a 7.5h/13GB trace and a 35h/48GB trace. Additionally a 1 hour trace with a total volume of 13GB from the network of a wireless provider used by about 1000 customers is used for the evaluation.

Figure 2 shows a schematic representation of the evaluation process. In the first step the collected trace files are processed by SPID and RT-ETD. The results of SPID are used as 100% baseline within the individual traffic categories. The filtered output files from the RT-ETD are then processed by SPID. The results are then compared to the results based on the original file. The metric we are using here is the fraction of flows in each category that can still be detected in the filtered file and the fraction by which the size of the trace file was reduced by the RT-ETD. An optimal result would be if still 100% of the encrypted flows are present in the filtered files, and no unencrypted flows are present.

Table 1 shows the results based on the ground truth traces. The worst results we get for encrypted eDonkeyTCP traffic, where 2.3% of the missing 5.5% is dropped due to a packet length of the first packet that is shorter than 16 bytes. An evaluation of including packets where the length is shorter than 16 bytes is left open for future work.

Based on the 1h/2GB real network trace we evaluated the usage for the coding-based classifier for TCP and UDP traffic. Using the coding-based classifier for UDP traffic does not influence the classification at all.

---

[1] Proxocket is a dll proxy for Winsock that dumps a copy of the network traffic to and from an application to a pcap file.
(http://aluigi.altervista.org/mytoolz.htm#proxocket)

For TCP traffic the coding-based classifier removes about 900kByte from the output file without changing the classification results of SPID. The 900kByte are plain POP3 flows.

Table 3 shows the results of the evaluation of using our algorithm as pre-filter for SPID. We are showing results for two traces collected in the cable provider network, and results for the trace from the WLAN provider network. An empty entry indicates a 0 count, this category is completely removed, whereas 0% indicates that the fraction is below 0.01%.

Between 73% and 96% of the flows are dropped by our pre-filter. Unencrypted flows such as FTP, HTTP, IMAP or SMTP are almost completely removed, which is a strong indication that only a small fraction of unencrypted traffic is forwarded. For encrypted protocols that we take into account, eDonkey TCP/UDP encrypted, MSE, Skype TCP/UDP at least 76.7% (MSE) of the flows are still present in the filtered file. For eDonkey and Skype the fraction is above 93%. SSH and SSL are detected as unencrypted due to the usage of a plain connection establishment. Such protocols can be easily detected by state of the art filtering methods and are thus outside of the scope of our work.

Table 3. Results for SPID pre-filter. The filter reduced the filesize by a factor of about 20. Well known encrypted protocols are removed, whereas a large fraction of encrypted flows is still present.

Type	7.5h/13GB original	filtered	35h/48GB original	filtered	1h/13GB original	filtered
Filesize [MB]	13531	3.36%	48527	2.94%	13157	12.5%
Sessions	242309	13.0%	1050206	4.37%	195243	27.1%
BitTorrent	64		8067		5061	
DNS	30946		91015		22166	
eDonkey					7169	
eDonkeyTCP encr.	9	100%	36	100%	9653	96.2%
eDonkeyUDP encr.	44	93.2%	95	100%	21808	96.0%
FTP	443		31		24	
HTTP	118226		210320	0%	46643	
IMAP	248		1861		26	
IRC			68		66	
ISAKMP	4		3	33%	227	
MSE	30	76.7%	280	99.3%	1323	81.7%
MSN	152		19		23	
POP	9770		25528		632	
SkypeTCP	773	99.5%	1167	99.4%	456	94.1%
SkypeUDP	18945	99.0%	27675	99.2%	6079	98.7%
SMTP	832		1075		53	
SpotifyServer	55	74.5%	90	94.4%	306	95.4%
SSH	946		60529		26	
SSL	10044		19203	0%	3210	
UNKNOWN	50778	23.3%	603144	2.8%	70292	21.2%

For the 1h/2GB real network we performed a detailed analysis on the filtered file. The trace is reduced to a size of 39.63MByte. Table 2 gives an overview of the categories that are still present in the filtered file. Almost 15% of the traffic is SYN+SYN/ACK traffic without any data communication. 63% of the traffic is encrypted traffic, where for the detection of the Skype traffic we are using the Adami Skype detector [11]. 19% of the traffic belong to a single flow where a binary request, invoked by a flash player, requests content from the AKAMAI distribution network.

7   Summary and Conclusions

In this paper we have presented how to use information from the first packet of a flow to identify encrypted traffic. The algorithm consists of two classifiers and can be used as real-time traffic filter as only the first packet of a flow has to be evaluated. The core classifier is based on payload entropy estimation, where entropy is used as a measure for uniformity which is an indication for encryption. The algorithm is refined by a further classifier, which takes into account the coding range used by the ASCII code. The main strength of the approach is its simplicity and accuracy. Evaluation based on encrypted ground truth traces and real-world network traces shows that more than 94% of the encrypted traffic is detected as encrypted, and more than 99% of the unencrypted traffic as unencrypted. Well known unencrypted protocols such as SMTP, HTTP, FTP, IMAP, POP3 and DNS are detected as unencrypted with probability as high as 99.9%.

A typical use case for our real-time traffic filter could be to pre-process data for L7 classifiers[2] where the focus is on encrypted flows, or detecting hidden traffic within encrypted flows. Using our approach the traffic volume that has to be handled by these classifiers can be reduced by a factor of about 10 to 50, depending on the traffic matrix.

References

1. Lyda, R., Hamrock, J.: Using entropy analysis to find encrypted and packed malware. IEEE Security & Privacy 5(2), 40–45 (2007)
2. Olivain, J., Goubault-Larrecq, J.: Detecting subverted cryptographic protocols by entropy checking. Research Report LSV-06-13, Laboratoire Spécification et Vérification, ENS Cachan (2006)
3. Pescape, A.: Entropy-based reduction of traffic data. IEEE Communications Letters 11(2), 191–193 (2007)
4. Dorfinger, P., Panholzer, G., Trammell, B., Pepe, T.: Entropy-based traffic filtering to support real-time Skype detection. In: IWCMC, Caen, France, pp. 747–751 (2010)
5. Shannon, C.E.: A mathematical theory of communication. Bell System Technical Journal 27, 379–423, 625–656 (1948)
6. Schürmann, T.: Bias analysis in entropy estimation. Journal of Physics A: Mathematical and General 37(27), L295–L301 (2004)
7. Paninski, L.: A coincidence-based test for uniformity given very sparsely sampled discrete data. IEEE Transactions on Information Theory 54(10), 4750–4755 (2008)
8. Paninski, L.: Estimation of entropy and mutual information. Neural Computation 15(6), 1191–1253 (2003)
9. Dorfinger, P.: Real-Time Detection of Encrypted Traffic based on Entropy Estimation. Master's thesis, Salzburg University of Applied Sciences, Austria (2010)
10. Hjelmvik, E., John, W.: Breaking and improving protocol obfuscation. Tech. Rep. 2010-05, Computer Science and Engineering, Chalmers University of Technology (2010), http://www.iis.se/docs/hjelmvik_breaking.pdf (28.01.2011)
11. Adami, D., Callegari, C., Giordano, S., Pagano, M., Pepe, T.: A Real-Time Algorithm for Skype Traffic Detection and Classification. In: Balandin, S., Moltchanov, D., Koucheryavy, Y. (eds.) ruSMART 2009. LNCS, vol. 5764, pp. 168–179. Springer, Heidelberg (2009)

---

[2] Classifiers mainly utilizing costly regular expressions on packet payloads.

Limits in the Bandwidth Exploitation in Passive Optical Networks Due to the Operating Systems (Poster)

Paolo Bolletta, Valerio Petricca, Sergio Pompei,
Luca Rea, Alessandro Valenti, and Francesco Matera

Fondazione Ugo Bordoni Viale del Policlinico 147, Rome, Italy
{pbolletta,spompei,lrea,avalenti,mat}@fub.it

Abstract. In this work we apply an accredited standard technique for end-user bandwidth evaluation in a wired access scenario and show limitations in the bandwidth exploitation of user of optical access networks due to the computer operating systems.

Keywords: bandwidth estimation, QoS, broadband access networks.

1 Introduction

Fiber To The x (FTTx, where the x can stand for Curb, Building or Home) accesses allow users to have ever increasing bandwidth, even though it has to be well understood if they will be able to exploit it in their broadband devices (PC, USB TV,...). One of the limitations in the exploitation of the access bandwidth is given by Operating Systems (OS) located in the broadband devices. In particular, an issue to be addressed regards on how much layer 2 bandwidth can be really exploited from upper OSI Layers. This topic has been already investigated in [1], and in [2] a detailed experimental investigation was shown for accesses based on xDSL, illustrating the impact of different OSes (Microsoft Windows XP, Microsoft Windows 7, Linux) on end-user bandwidth estimation; moreover, results showed how such differences were stronger at higher bit-rates. Therefore, we foresee that such differences will be wider and wider using optical accesses in FTTx networks.

In this work, we refer to a bandwidth estimation method proposed by European Telecommunications Standards Institute (ETSI) in [3] and we report an experimental investigation about the role of OSes on the effectiveness of such bandwidth estimation technique for Gigabit Passive Optical Network (GPON) [4], *i.e.* the fiber access technology preferred by many operators.

The extended abstract is structured in the following way: after a description on the methods to measure the user Quality of Service (QoS) in Section 2, in Section 3 we report the experimental set-up and in Section 4 results for different user bit-rates are shown according to three well known OSes.

J. Domingo-Pascual, Y. Shavitt, and S. Uhlig (Eds.): TMA 2011, LNCS 6613, pp. 172–175, 2011.
© Springer-Verlag Berlin Heidelberg 2011

2   Bandwidth Exploitation and Measurement in Broadband Access Networks

Today's bandwidth estimation plays a key role in telecommunication evolution and market regulation in order to correctly define service level agreement between customers and Internet Service Providers (ISP). On the other hand customers are interested in network performance verification to better exploit ISP market competition. For these reasons this study aims for investigating bandwidth estimation techniques based on active probes designed considering protocols commonly used by customers: in this way bandwidth estimation matches the real experienced performance.

In this work, we look at the evaluation of protocol performance and at the impact of software implementation, considering current QoS evaluation best practice and looking forward at the Next Generation Access Networks (NGAN). Our method was already used in different access network scenarios, in terms of access bandwidth and network delay. In particular, in [2, 5] we reported measurements regarding ADSL2+ that is the most adopted technology by Telecommunication Operators in Europe. First of all, we consider a specific QoS evaluation technique, based on the ETSI EG 202 057 [3], using File Transfer Protocol (FTP) [6] probes. In such assumption, the Transmission Control Protocol (TCP) plays a key role in evaluating performance, since it directly regulates the data flow. The choice of a TCP dependent technique tries to keep QoS evaluation as closer as possible to the end-user effective experience of broadband access services.

3   Experimental Setup

The network test-bed is shown in 1: the core part is composed of four Juniper M10 (J1, J2, J3, J4) routers fully meshed using the fibers deployed in the cable Roma-Pomezia-Roma (50 Km with round trip in Pomezia). Server and GPON Optical Line Termination (OLT) are connected to the core network by means of two edge routers (Cisco 1, Cisco 2) using fiber Gigabit Ethernet transmission.

To establish an End-to-End controlled bandwidth, we use the technique described in [7] that allows us to assign a guaranteed bandwidth between two end-points of the network by means of different tagging techniques, *i.e.* Virtual LAN and Virtual Private LAN Service (VPLS).

In this context, the only parameter taken into account is the network delay, in order to emulate a wide set of network topologies and conditions in terms of geographical extension (simplified as network delay). To achieve this goal, a delay generator is used to obtain different Round Trip Times (RTTs) between Server and Client. In such tests, a FTP-based QoS estimation tool is implemented in the three most popular OSes: Windows XP SP3, Windows 7, Linux Ubuntu 10.4. The aim is to outline the impact of TCP implementation in QoS evaluation and bandwidth customer exploitation.

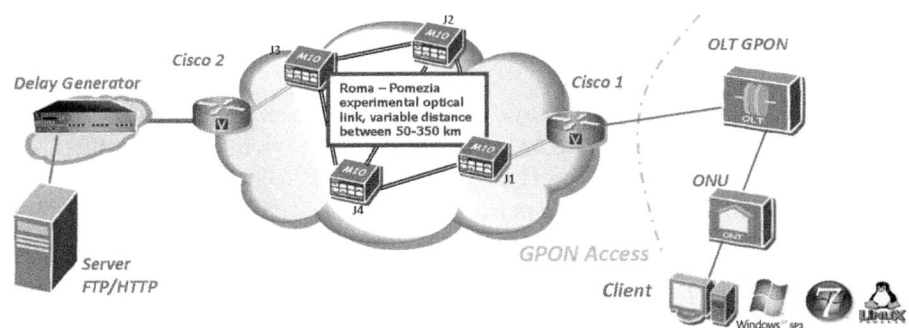

Fig. 1. Test-bed architecture

4   Results

The FTP-based bandwidth evaluation method, as used in our testing activity, measures FTP Throughput considering data flow between FTP Server process and FTP Client process located in different network portions, respectively provider side and customer side. FTP sessions make possible to estimate data transmission rate and the transmission delay experimented by downloading and/or uploading specified test files several times (50 repetitions) between a remote site and a user's terminal.

In Fig.2 experimental results are reported, average throughput values are represented vs. network delays. We considered two typical bandwidth profiles. In the first one, we considered GPON with 128 users that can provide a physical bandwidth of about 18 Mbit/s. In the second one, a GPON with up to 32 users was used, with a physical bandwidth of about 100 Mbit/s.

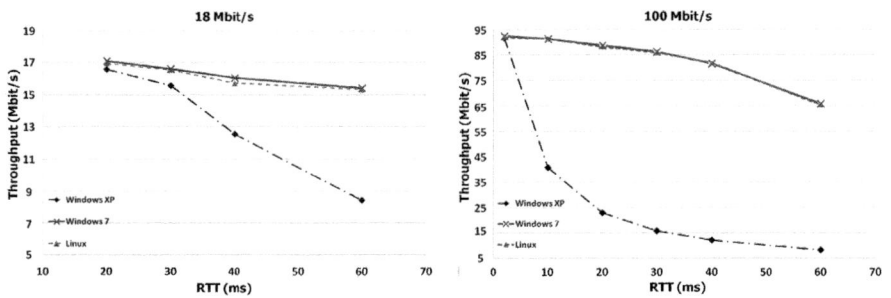

Fig. 2. TCP connection Throughput evaluation

Like other window-based protocols, such as TCP, performance depend on RTT; in particular, these results point out that, for specified access conditions, OSes in client host have a considerable impact on FTP performance.

Results reported in Fig. 2 outline the critical impact of network delay and increasing of bit-rates considering TCP protocol implemented in MS Windows XP. In particular, we noticed a reduction of throughput, with respect to the nominal one, equal to 50% for 18 Mbit/s profile, and equal to 90% for 100 Mbit/s profile. Evident substantial differences (Fig. 2) between Win XP and advanced OSes (Win 7 and Linux) are due to enhanced TCP algorithm implementation, related to adaptive parameters in the algorithm (*i.e.* Auto-Tuning of receiver windows size [8]). Furthermore, it is important to remark that a performance degradation can be observed also considering advanced OSes for high values of network delay. For example, for 100 Mbit/s profile, only 65 Mbit/s are measured by QoS measurement tests when RTT is 60 ms.

5   Conclusion

In this work, we have reported an experimental investigation about the role of Operating Systems on the bandwidth exploited by GPON users. The results reported in this paper show that the dependence of the QoS on the Operating Systems increases with user bit-rates, and how effectiveness of the bandwidth estimation is affected by considering broadband Optical Access. In particular, we noticed a reduction of throughput, with respect to the nominal one, up to 50% for 18 Mbit/s profile, and up to 90% for 100 Mbit/s profile. Therefore, it means that users could be strongly limited in the exploitation of the bandwidth, and such limitation is much relevant in case of optical access networks that should permit very wide bandwidth. As a consequence, we believe that a testing scenario for FTT*x* accesses needs to be described not only according to the physical parameters, but also paying attention to software implementation factors that could affect the testing results.

References

1. Jansen, S., McGregor, A.: Measured comparative performance of TCP stacks. In: Dovrolis, C. (ed.) PAM 2005. LNCS, vol. 3431, pp. 329–332. Springer, Heidelberg (2005)
2. Del Grosso, A., et al.: On the Impact of Operative Systems Choice in End-user Bandwidth Evaluation: Testing and Analysis in a Metro-access Network. In: IARIA ACCESS 2010, Valencia, Spain (September 2010)
3. ETSI EG 202 057-4 User related QoS parameter definitions and measurement. European Telecommunications Standards Institute (October 2005), http://www.etsi.org
4. ITU-T G.984.1, Gigabit-capable Passive Optical Network (GPON) (March 2003)
5. Bolletta, P., et al.: Monitoring of the User Quality of Service: Network Architecture for Measurements and role of Operating System with consequences for optical accesses. In: ONDM 2011, Bologna, Italy (February 2011)
6. Postel, J., Reynolds, J.: File Transfer Protocol. STD 9, RFC 959 (October 1985)
7. Valenti, A., et al.: Quality of Service control in Ethernet Passive Optical Networks based on Virtual Private LAN Service. IET Electronics Letters 45(19), 992–993 (2009)
8. Jacobson, V., et al.: TCP Extensions for High Performance. RFC 1323 (May 1992)

Toward a Scalable and Collaborative Network Monitoring Overlay (Poster)

Vasco Castro, Paulo Carvalho, and Solange Rito Lima

University of Minho, Department of Informatics,
4710-057 Braga, Portugal
{pmc,solange}@di.uminho.pt

Abstract. This paper presents ongoing work toward the definition of a new network monitoring model which resorts to a cooperative interaction among measurement entities to monitor the quality of network services. Exploring (i) the definition of representative measurement points to form a network monitoring overlay; (ii) the removal of measurement redundancy through composition of metrics; and (iii) a simple active measurement methodology, the proposed model aims to contribute to a scalable, robust and reliable end-to-end monitoring. Besides the model proposal, a JAVA prototype was implemented to test the conceptual model and its design goals.

1 Introduction

Monitoring of large networks raises multiple challenges regarding scalability, robustness and reliability of measurements. It is known that monitoring systems where a single point is responsible for gathering and processing measurements obtained throughout the network suffer from severe scalability and robustness limitations. To address this problem, distributed solutions where monitoring data is collected and processed at each measurement point (MP) have been proposed. For instance, solutions based on active edge-to-edge measurements provide a straightforward way of measuring service quality, however, the potential interference of cross probing among boundary nodes on network behaviour needs to be carefully considered.

To reduce network overhead and improve spatial coverage, it is important to identify the most representative and critical network points in order to obtain an overall view of the network status involving only a subset of MPs. Resorting to composition of metrics between these MPs, i.e. through concatenation of partial metrics, the interference on network operation can be reduced, avoiding redundant measurements in overlapping links. The composition of metrics also allows observing trends, being more informative as a result of the underlying metric partitioning scheme.

In this context, this paper proposes a collaborative network monitoring overlay which resorts to the cooperation between representative MPs strategically located in the network to compute performance and quality metrics both intra-area and end-to-end. The aim is to pursue a flexible, scalable and accurate monitoring overlay solution that simplifies and systematises the cumulative computation of metrics by involving a subset of network nodes.

J. Domingo-Pascual, Y. Shavitt, and S. Uhlig (Eds.): TMA 2011, LNCS 6613, pp. 176–180, 2011.
© Springer-Verlag Berlin Heidelberg 2011

This paper is organised as follows: related work is discussed in Section 2, the proposed monitoring model and its components are described in Section 3, the model prototype is presented in Section 4 and the conclusions are summarised in Section 5.

2   Related Work

Active monitoring carried out on an edge-to-edge basis, i.e., between network boundaries, is particularly suitable for monitoring network performance and quality of service (QoS) [1]. This approach improves scalability by involving only edge nodes in the monitoring process, removing the complexity of monitoring tasks from the network core. Considering that in edge-to-edge probing, probes from distinct pair of edges may cross the same links, hop-to-hop monitoring strategies try to avoid repeating probes in those links. However, capturing network behaviour combining hop-by-hop measures is not an efficient and easy solution as it involves : (i) a high-degree of metrics' concatenation; (ii) monitoring agents in all network nodes; and (iii) additional traffic in the network for reporting metrics to management stations. To reduce the amount of data exchanged between management stations and MPs, several solutions have been pointed out, namely the use of flow aggregation [2], statistical summarisation [3] and network thresholds crossing alerts [4].

Tomography concepts [5] continue to deserve significant attention for estimating distinct aspects of network behaviour, including QoS [6,7]. In [8], network tomography is applied to the definition of a monitoring overlay to infer packet loss in all network nodes.

Taking in consideration the mentioned strategies, this study proposes a network monitoring overlay solution which resorts to representative MPs to compute performance and quality metrics both intra- and inter-area, with reduced overhead.

3   A Collaborative Monitoring Overlay

The proposed model relies on a collaborative participation of representative MPs acting as peers, each one contributing with a disjoint measure component to the evaluation of the global measure. Thus, end-to-end measurements are obtained through the aggregation of metrics calculated in each of the network areas involved.

Figure 1 illustrates the monitoring overlay network and the underlying physical topology. The overlay network consists of representative MPs and these are the only players taking part in the measurement process. Each MP in the overlay is expected to store the measurements to its neighbouring MPs. Thus, measurement data is distributed and stored throughout the overlay network. Based on a monitoring request, each MP in the measurement path provides the required measures for aggregation in order to calculate a set of metrics between any specified MPs. Distributing measurement data over several MPs also enables a rapid recovery of the measurement process by bringing alternative MPs in the process of rebuilding the measurement path in case of routing or network topology changes. Note that these changes do not necessarily imply a change in the overlay topology.

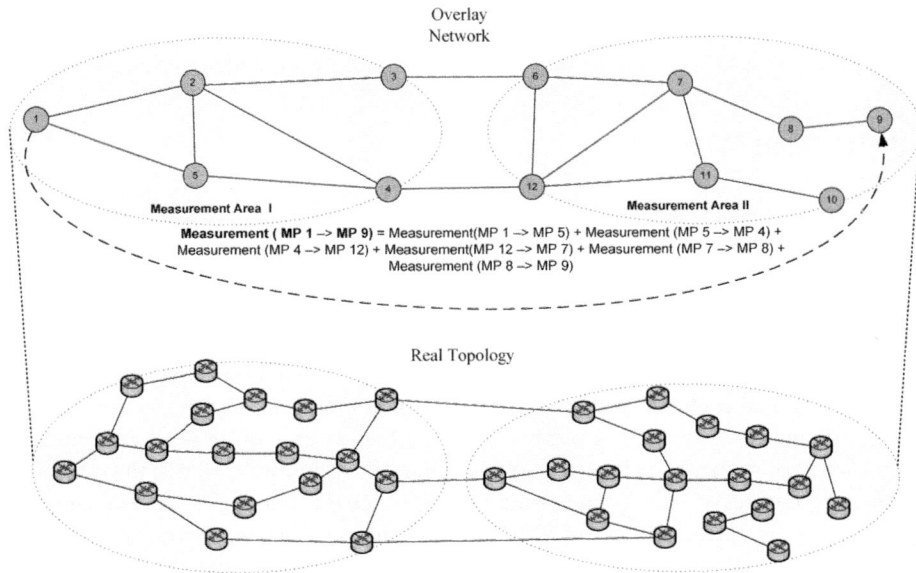

Fig. 1. Example of measurement between different administrative areas

The proposed model allows measurements at two levels: Intra-area and Inter-area. *Intra-area* measurements are carried out on a regular time basis to ensure that MPs in the same area have a clear view of network status and quality of service. A MP may, at anytime, send or exchange measurement data between itself and any other MP within its area. Thus, by retrieving data from multiple MPs in the area and using composition of metrics, it is possible to calculate the value of a metric for a given measurement path. *Inter-area* measurement is performed through the composition of the metrics resulting from intra-area measurements. Conversely to intra-area operation, this type of measurement does not need to be performed continuously, but on request. This process can be triggered, for example, by an application signalling process to assess the communication path before establishing an end-to-end session crossing different network areas.

Model operation - Initially, a monitoring entity sends a message to the initial MP indicating that it needs to obtain a set of metrics between a pair of MPs (see example in Figure 1). This MP, after receiving the request, sends it in the overlay network as a packet measurement request. Each MP in the overlay path will intercept this packet and attach measurement data between itself and the upstream MP, before sending it to the downstream MP. This process is repeated until the destination is reached, i.e., each MP will successively attach its measurement data along the overlay. The final MP or the destination, upon receiving the packet measurement request, will proceed similarly, sending subsequently the resulting message back to the initial MP with all collected measurements. At this point, the initial MP is able to compose the required metrics in order to obtain end-to-end measures. This operation can assume distinct cumulative functions (additive, multiplicative, max-min, etc.) depending on the nature of the metric being evaluated.

In practice, this process can be considerably simplified as area border MPs (e.g. MP 12 in Figure 1) may already have up-to-date measurements from the remaining measurement path. This allows an immediate reply from that MP to the measurement requester, reducing measuring latency significantly. This process can be further improved through proper pro-active metrics dissemination among inter-area MPs.

One challenge of the present model is to identify the representative MPs. Although several works target this topic [8,9], this aspect requires further study. Other relevant aspect currently under study is focused on solutions to avoid measurement data fragmentation.

4   The Implemented Prototype

The model prototype was implemented in Java and MySQL, being the measurement primitives structured in XML. The implementation includes four main components: (i) the "Measure Requester"; (ii) the "Packet Interceptor"; (iii) the "Measure Processor"; and (iv) the "Measure Receiver", interacting as illustrated in Figure 2.

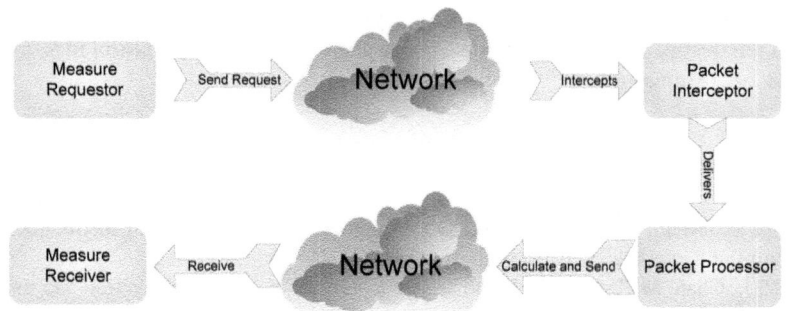

Fig. 2. Interaction of Model Components

Measure Requester - This component is responsible for initiating the measurement process between two MPs. In the developed prototype, this is a command line application that receives as parameters, the source and destination MP, and the set of metrics to measure.

Packet Interceptor - This component is responsible for capturing measurement packets. These packets are differentiated in the network through the use of `router alert` option within IPv4 header, avoiding packet processing at upper protocol layers. In a Linux router, this can be accomplished resorting to proper `iptables` packet filtering (it requires the extension `xtables-addons`). Captured packets are taken from kernel to user space (through `libnetfilter_queue`) for processing at MPs. The use of `router alert` option avoids the use of explicit MP addressing, allowing for a more flexible overlay topology definition.

Packet Processor - This component is responsible for processing and concatenating measurement data. Once a packet request is intercepted at an MP, this component detects the new request, validates it and appends the required metrics to the measurement packet. This process involves identifying the latter upstream MP before adding its measurement contribution. Then, the component builds an IP packet setting the `router alert` option, updates the data payload accordingly and sends the packet to the downstream MP. Once the last MP is reached, the "Packet Processor" opens a TCP connection to the initial MP for sending the aggregate measurement outcome.

Measure Receiver - When the measurement process starts, a measurement packet request is issued and, simultaneously, the request is stored in a database, remaining in listening mode on an UDP port. Upon receiving the corresponding measurement result, this component updates the database for the corresponding request.

5   Conclusions

This paper has presented ongoing work toward the definition of a network monitoring overlay which resorts to a cooperative interaction among representative MPs to monitor the quality of network services. In the proposed model, measurement overhead and redundancy are reduced through the composition of metrics from non-overlapping measurement paths, both intra- and inter-area. This aspect along with the ability to accommodate network topology changes aim to contribute to a scalable and flexible end-to-end monitoring solution. A JAVA prototype has also been implemented to test the conceptual design goals of the model, being currently under evaluation in a virtualised network environment.

References

1. Habib, A., Khan, M., Bhargava, B.: Edge-to-edge measurement-based distributed network monitoring. Computer Networks 44, 211–233 (2004)
2. Lin, Y.J., Chan, M.C.: A scalable monitoring approach based on aggregation and refinement. IEEE JSAC 20 (2002)
3. Asgari, A.H., Egan, R., Trimintzios, P., Pavlou, G.: Scalable monitoring support for resource management and service assurance. IEEE Network 18(6), 6–18 (2004)
4. Wuhib, F., Stadler, R., Clemm, A.: Decentralized service-level monitoring using network threshold crossing alerts. Communications Magazine 44(10), 70–76 (2006)
5. Vardi, Y.: Network Tomography: Estimating Source-Destination Traffic Intensities from Link Data. Journal of the American Statistical Association 91(433), 365–377 (1996)
6. Gu, Y., Jiang, G., Singh, V., Zhang, Y.: Optimal probing for unicast network delay tomography. In: INFOCOM 2010, pp. 1244–1252. IEEE Press, Piscataway (2010)
7. Arya, V., Duffield, N., Veitch, D.: Temporal Delay Tomography. In: INFOCOM, pp. 276–280 (2008)
8. Chen, Y., Bindel, D., Katz, Y.H.: Tomography-based Overlay Network Monitoring. In: ACM SIGCOMM Internet Measurement Conference (IMC), pp. 216–231. ACM Press, New York (2003)
9. Ratnasamy, S., Handley, M., Karp, R.M., Shenker, S.: Topologically-Aware Overlay Construction and Server Selection. In: INFOCOM (2002)

Hardware-Based "on-the-fly" Per-flow Scan Detector Pre-filter (Poster)

Salvatore Pontarelli, Simone Teofili, and Giuseppe Bianchi

Università degli Studi di Roma Tor Vergata
Via del Politecnico, 1 00133 Rome, Italy
{salvatore.pontarelli,simone.teofili,giuseppe.bianchi}@uniroma2.it

Abstract. Pre-filtering monitoring tasks, directly running over traffic probes, may accomplish a significant degree of data reduction by isolating a relatively small number of flows (likely to be of interest for the monitoring application) from the rest of the traffic. As these filtering mechanisms are conveniently run as close as possible to the data gathering devices (traffic probes), and must scale to multi-gigabit speed, the feasibility of their implementation in hardware is a key requirement. In this paper, we document a hardware FPGA implementation of a recently proposed network scan pre-filter. It leverages processing stages based on Bloom filters and Counting Bloom Filters, and it is devised to detect, through on-the-fly per-packet analysis, the flows which potentially exhibit a network/port scanning behaviour. The framework has been implemented in a modular manner. It suitably combines two different general-purpose modules (a rate meter and a variation detector) likely to be reused as building blocks for other monitoring tasks. In the following presentation, we further discuss some lessons learned and general implementation guidelines which emerge when the goal is to efficiently implement run-time updated (*i.e.*, dynamic) Bloom-filter-based data structures in hardware.[1]

1 Introduction

In high-speed (multi-gigabit) networks, traffic analysis and network monitoring functions based on the traditional gather-first-process-later paradigm appear inconvenient. Indeed, the relative number of flows which exhibit a behaviour worth of detailed investigation can be small with respect to the total amount of traffic carried over a network link, and monitoring solutions which mandate to deliver *all* the captured traffic to a remote device for analysis and inspection appear to pose a huge and unnecessary demand on the network monitoring infrastructure.

Starting from the seminal paper from Estan and Varghese [1], a new direction in traffic monitoring emerged. The idea is to delegate some traffic processing tasks to traffic probes. Specifically, probes may support traffic analysis functions whose goal is to detect and isolate flows which exhibit a potentially anomalous

---

[1] This work has been partially supported by the European Commission in the frame of the DEMONS project http://fp7-demons.eu/

J. Domingo-Pascual, Y. Shavitt, and S. Uhlig (Eds.): TMA 2011, LNCS 6613, pp. 181–184, 2011.
© Springer-Verlag Berlin Heidelberg 2011

behaviour, where said "anomalous behaviour" is suitably defined in correspondence of the specific monitoring application goals. As a result, monitoring infrastructure scalability is accomplished by dramatically reducing the amount of data ultimately delivered to remote monitoring applications.

An extensive amount of literature has shown that several meaningful traffic analysis tasks, such as heavy flows detection [1], traffic classification [2], worm fingerprinting [3], scan detection [4][5], and so on, can be conveniently performed "on-the-fly" over a memory/resource-constrained device such as a hardware traffic probe. When such analysis function are implemented as pre-filters, the accuracy requirements are somewhat reduced (the goal to provide a final answer is delegated to the remote monitoring application), and approximate low-resource consuming data filters and structures, such as those based on Bloom Filters and their extensions [6], seems very appealing.

Among these papers, our previous work [5] presents a novel method for detecting and accounting for *variations*, which are usually a component of a more comprehensive network scan activity. Specifically, we aim at detecting flows which exhibit a significant difference, in time, with respect to one or more parameters (for instance traffic which originates by a same IP source address and "hits" a large number of IP destinations; ARP requests showing a large IP target variation; TCP packets addressed to a large amount of ports, etc). It was shown that this approach may achieve a significant reduction ($95 - 98\%$) in terms of number of flows which require further fine-grained analysis, meanwhile using a relatively small amount of memory resources over the traffic probe (order of a few hundreds of kbits, dependent on the accuracy target).

This work complements our former work [5] by showing how the specific filtering mechanism therein proposed can be implemented on a reprogrammable hardware device (*i.e.* FPGA), thus proving the the viability of that approach in a high speed network monitoring scenario.

2   Architecture

Figure 1 shows the two-stage filtering approach proposed in [5]. For reasons of space, the reader is referred to the original paper for a detailed functional explanation. In what follows, we focus on the hardware implementation decisions concerning each stage, indeed influenced by the need to handle Bloom-type filters dynamically updated on a per-packet basis.

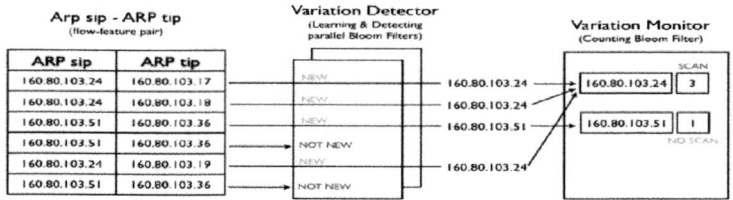

Fig. 1. Two stage filtering - example for ARP traffic

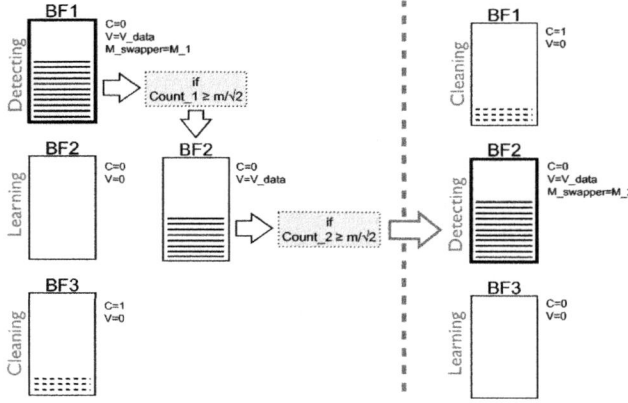

Fig. 2. variation detector: Swapping Bloom Filters

2.1   Stage 1: Variation Detector

We recall from [5] that a variation detector can be constructed using two self-clocked swapping Bloom Filters. The major hardware implementation concern we had to face consisted in the time needed to reset one of the two Bloom filters upon filter swapping. Indeed, such reset procedure needed to be accomplished "instantaneously", *i.e.* in the short time frame elapsing between the arrival of two consecutive packets.

As a solution, we have resorted to implement three Bloom filters instead of the original two. Such three Bloom filters are in three different operating states: Detecting, Learning, Cleaning. The filter in Detecting state, for each received packet, checks if the string obtained by combining the selected flow key (e.g. the IP source address for network scan) and the selected feature key which is monitored to identify a variation in the flow (e.g. the IP destination address) is already stored in the Bloom Filter; if not, it is inserted and the flow is accounted in the next variation monitor module. The filter in Learning state is updated according to the same rule, but its output is discarded. The only role of the Cleaning state is to wipe the content of the filter. Specifically, when the Detecting filter is full (according to the rules defined in [5]), filter swapping occurs so that the previously Learning filter becomes the Detecting one, and this latter enters into Cleaning state.

Note that the development of a supplementary Cleaning filter was a necessary consequence of our design decision of developing Bloom filters in the FPGA Block RAM: RAM zeroing in fact requires a number of clock cicles linearly increasing with the filter dimension, and in any case always greater (for practical filter sizes) than the time available between two consecutive packets. Specifically we need a clock cicle to reset 32 bits (*i.e.* 1000 clock cicles for Bloom filter of 32000 bits). The alternative solution of implementing the Bloom filters in LUTs was not considered viable because of FPGA area consumption. The implementation of 3 Bloom Filters of 32000 bits each would have required the usage of 93% of the logic resources of a Xilinx Virtex 5 [7].

2.2   Stage 2: Variation Monitor

The variation monitor is composed of a Counting Bloom Filter (CBF) Block that receives, as inputs, the flow string signalled by the variation detector as carrying a new flow/feature pair. Such CBF is transformed into a rate monitor [5] by periodically decrementing all the filter bins. Again, this specific additional functionality requires a careful hardware implementation, when (as per our choice), also the CBF is implemented using the Block RAM. Note that a periodic decrement operation, for the same reasons described above, applied to all the CBF bins requires a large number of memory accesses (further slowed by the need to perform arithmetic operations over the bin values themselves). However, in this case, there is no possibility to deploy an associated cleaning filter; rather, the hardware must make possible to decrement the filter while its state is maintained and eventually updated. As a result,we have introduced a pointer to the last decremented counter together with a compensation circuitry subtracting 1 to the value taken from the bins not yet decremented.

3   Results

The presented paper proposes an implementation of the framework presented in [5] on a reprogrammable hardware device (*i.e.* FPGA). Our single-pipe implementation, on a low-end Virtex II pro [8] reaches the frequency of $183MHz$ overcoming the minimum frequency required to process data over a $1Gb/sec$ link ($125MHz$). The achievement of higher rates (in the order of few tens of Gbps) appears easy by 1) exploiting more performing FPGA, and 2) by designing multiple parallel processing pipe. The implemented framework exploits a very limited amount of physical resources. In a Virtex II pro [8] that is a low-end FPGA we need only 20% of the Block RAM (50 of 232) and the 4% of the Flip-Flop (1920 of 46632). In a Virtex 5 [7] we exploit 15% of the Block RAM and less then 2% of the Flip-Flop.

References

1. Estan, C., Varghese, G.: New Directions in Traffic Measurement and Accounting. In: SIGCOMM 2002 (August 2002)
2. Dharmapurikar, S., Song, H., Turner, J., Lockwood, J.: Fast Packet Classification Using Bloom Filters. In: Proceedings of the 2006 ACM Symposim on Architectures for Networking and Communications Systems (December 2006)
3. Singh, S., Estan, C., Varghese, G., Savage, S.: Automated Worm Fingerprinting. In: 6th Usenix Symposium on Operating Systems and their Applications (2004)
4. Kong, S., He, T., Shao, X., Li, X.: A Double-Filter Structure Based Scheme for Scalable Port Scan Detection. In: IEEE ICC 2006, Istanbul, Turkey (2006)
5. Bianchi, G., Boschi, E., Teofili, S., Trammell, B.: Measurement Data Reduction through Variation Rate Metering. In: INFOCOM 2010 (March 2010)
6. Broder, A., Mitzenmacher, M.: Network Applications of Bloom Filters: A Survey. Internet Mathematics 1(4), 485–509 (2005)
7. Virtex-5 Family Overview, http://www.xilinx.com/support/documentation/data_sheets/ds100.pdf
8. Virtex-II Pro and Virtex-II ProX FPGA User Guide, http://www.xilinx.com/support/documentation/user_guides/ug012.pdf

Anti-evasion Technique for Packet Based Pre-filtering for Network Intrusion Detection Systems (Poster)

Salvatore Pontarelli and Simone Teofili

Consorzio Nazionale InterUniversitario per le Telecomunicazioni (CNIT)
University of Rome "Tor Vergata", Via del Politecnico 1,
00133, Rome, Italy

Abstract. This work proposes a method to extend packet pre-filtering for Network Intrusion Detection Systems (NIDS). The aim of the method is to avoid the false negatives occurring when a malicious content has been sent splitted in several packets. In this paper we propose a method that is able to identify even the fragmented malicious content avoiding false negative limiting the false positive rate.[1]

1 Introduction

Network Intrusion Detection Systems (NIDS) like Snort [1] are used for monitoring the presence of different kind of attacks in a network. The packets or flows carrying these attacks are detected looking for in the header and payload "malicious content" defined by specific IDS rules. IDS software need high computational resources and processing time to examine all the packets of a network. These problems strongly limit the usage of IDS in the backbone of Internet Service Provider network. The porting of all the components of a software based IDS to an hardware platform is not feasible. Instead it is possible to identify some tasks that form one side allow to offload the software IDS and from the other can be easily implemented in hardware. Starting from this consideration hardware based packet pre-filtering that allow inspection at wire speed has been proposed in [2]. The main limitation of this approach is that the it does not perform any reordering of the packets composing the data stream. The absence of reordering causes a false negative (i.e. the flow contained an attack is not identified) when the malicious contents are transmitted in different packets. The obvious solution to avoid these false negatives is to implement in hardware the TCP reassembly and IP defragmentation tasks [3]. In Snort [1] this tasks are accomplished by STREAM5 and Frag3 pre-processors.

The implementation of this task is,however, very complex, and the papers that describe hardware implementation of TCP/IP reassembly usually can cope with a very limited number of flows. The solution we propose in this paper

---

[1] This work has been partially supported by the European Commission in the frame of the DEMONS project http://fp7-demons.eu/

J. Domingo-Pascual, Y. Shavitt, and S. Uhlig (Eds.): TMA 2011, LNCS 6613, pp. 185–188, 2011.
© Springer-Verlag Berlin Heidelberg 2011

overcomes the limitation described above preserving a per-packet anomalous content analysis that allows us to manage the thousand of flow travelling into a network. Our solution is based on searching also for half-content and limits the false positive rate by selecting a best content representing each rule.

2   Proposed Strategy

In this section we show how to modify the IDS rules to avoids false negative due to fragmentation. First of all we apply the method presented in [4] to simplify the inspected rule taking the most representative content of each rule. This choice allows to limit the false positive rate due to the use of a relaxed version of the IDS rules. In Fig. 1 is shown how a content can be split between two packets. We suppose that the content is at least long L bytes, while the packets whose payload is longer than L. Three cases can occur:

1. the content is contained in a single packet (Fig. 1 a)
2. the content is equally divided between two packets (Fig. 1 b)
3. one packet contains more than $L/2$ bytes, the other less than $L/2$ (Fig. 1 c).

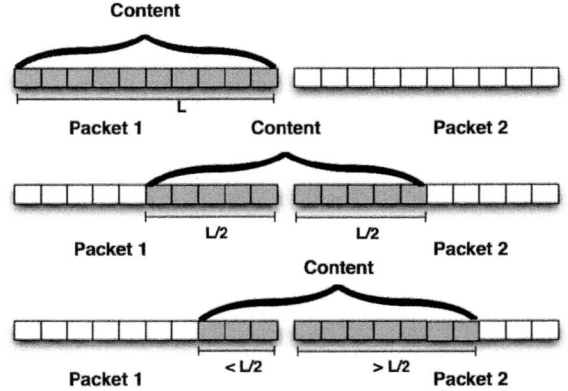

Fig. 1. Different split of a content in consecutive packets

The figure shows that, supposing all the packets longer than L, one of the packet containing a splitted content has to contain at least $L/2$ bytes. Moreover we note that half of the malicious content is contained always in the last $L - 1$ bytes of a packet, or in the first $L - 1$.

To check if a content is present in a flow it is sufficient to check the presence of the first half part of the content in the last $L - 1$ bytes of the packet or the presence of the second half part of the content in the first $L - 1$ bytes of a packet. If one of these checks is successful it is possible that a suspicious content has been sent into the network by using consecutive packets. Forwarding the flow that contains this packet to the software IDS we are therefore able to avoid false negative due to fragmented packets. We notice that this method works only under the assumption that all the packets had at least a length of L bytes. In fact, suppose to have a packet with length of $L - 2$ bytes, an attacker can split a content L into three packets. This fragmentation can evade the presented technique. To avoid the problem, we propose to forward to the software IDS the flows that had a packet with length less than L for further inspection.

3   Experimental Result

This section present the results obtained by applying the method described above to a realistic network scenario. In this experiment, we fix the parameter L to twenty and therefore $L/2$ is set to 10 bytes. In order to validate our results we select a large subset of the available snort IDS rules [1] to evaluate the false positive rate of our methodology. The rules has been used to analyze about 700 MBytes of traffic collected on the local area network of the Tor Vergata University network group. The collected trace is composed by a million of packet belonging to about 13000 different flows. The rules have been modified in order to match also the half-content and the begin or at the end of a packet, as described in the previous section. The packet distribution length is presented in Table 1.

Table 1. Distribution of packet length and number of flow with small packets

packet length (L)	number of packets	(%)	number of flows	(%)
< 100	10462	1%	1862	15 %
< 50	7606	0.7 %	1291	10 %
< 20	2786	0.2 %	842	7 %

From the above table we estimate that the 7% of the traffic had a packet with a length that can not be properly managed by our half-content method. Finally, we further reduce this amount of flow considering that many of these flows had as small packet the last packet of the stream, (if the small packet is the last one the content can not be split in three packets). Fig. 2 depict how a content can be split between these two packets.

It can be seen that in the second to last packet at least half-content is stored. Considering this refinement the number of flow that should be forwarded to the software decrease to 286 (the 2%), because 556 flows had the last packet with length less than L bytes. In Table 2 the result of our experiment are presented. The reassembly column use the original set of rules, while the half-content column use our method.

All the flow detected as suspected (the 507 flows of Table 2) and the ones composed with small packets (286 flows) are sent to an IDS with reassembly for further analysis. We obtain a total of 793 flows that are detected as suspected by our method again the Snort IDS that identify only 287 flows. Our method therefore identifies as flows needing further analysis about 6% (793 flows over a total of 13000 flows) of the incoming flows, offloading the software from the analysis of the overall flow travelling into the network.

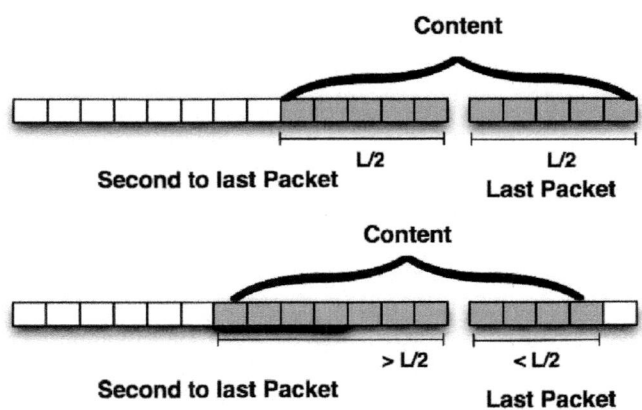

Fig. 2. Different split of a content in second to last and last packets

Table 2. Alert generated

	Snort with reassembly	proposed pre-filtering
inspected flows	13000	13000
flow sent to Snort	-	793
flow detected by Snort	287	287
Lost attacks	-	0
traffic reduction	-	7%

4   Conclusions

In this paper has been presented a methodology that avoids false negative due to packet fragmentation in packet pre-filtering for NIDS. The experiments performed on real traffic case are presented in order to prove the effectiveness of our proposed solution. Our measurement shows that our method forwards only the 6% of the incoming traffic to the second stage of the NIDS.

References

1. Sourcefire, Snort: The Open Source Network Intrusion Detection System (2003), http://www.snort.org
2. Song, H., Sproull, T., Attig, M., Lockwood, J.: Snort offloader: a reconfigurable hardware NIDS filter. In: International Conference on Field Programmable Logic and Applications, August 24-26 (2005)
3. Necker, M., Contis, D., Schimmel, D.: TCP-Stream Reassembly and State Tracking in Hardware. In: 10th Annual IEEE Symposium on Field-Programmable Custom Computing Machines, FCCM 2002 (2002)
4. Teofili, S., Nobile, E., Pontarelli, S., Bianchi, G.: Snort pre-filter for data-reduced intrusion detection: hardware design issues and trade-offs. In: International Tyrrhenian Workshop on Digital Communications (ITWDC 2010), Ponza, Italy, September 6-8 (2010)

COMINDIS – COllaborative Monitoring with MINimum DISclosure (Poster)

Jacopo Cesareo[1], Andreas Berger[2], and Alessandro D'Alconzo[2]

[1] Princeton University
jcesareo@princeton.edu
[2] FTW Telecommunications Research Center Vienna
{berger,dalconzo}@ftw.at

Motivation: In the modern Internet, network anomalies are manifold and range from Distributed Denial of Service (DDoS) attacks over unsolicited communication (e.g. Spam), to large-scale information harvesting. Network operators react by deploying carefully selected monitoring equipment, tuned to protect their individual core assets. Consequently, there exist a multitude of different views on the activities of a particular host at one moment in time, depending on the locally observed activity patterns, the configurations of the monitoring equipment, and the policies and legislations which influence the amount of traffic information that can be analyzed.

Collaborative monitoring approaches try to benefit from sharing this diverse information amongst operators. The challenge thereby is to share enough data while preserving confidential or business-sensitive information. Approaches based on secure multiparty computation (as e.g. SEPIA [2]) address this problem by running computations on encrypted data, thereby avoiding to disclose any confidential data. However, this comes at a high computation cost and can therefore not be used for real-time monitoring tasks that involve vast amounts of data. Other systems (like e.g. [1]) try to avoid leaking sensitive information by mangling it before sharing it with other parties. These anonmyization operations, in addition to complicating the analysis of the traffic data, implicitly strip all context information, like monitoring location, configuration parameters of the detection device, the overall network "health" (i.e., other local network problems at the same time that could be related?), and specific network characteristics in normal conditions. In this paper, we assume unavailability of detailed traffic information, but consider a more realistic setting where operators share notions of suspiciousness regarding particular Internet hosts. In addition to the principle of minimum disclosure, we show that this approach has a number of advantages. Particularly, it leverages the fact that each operator knows its network best, and can thus give the most reliable verdict about the current activities.

System Overview: COMINDIS stores, maintains, and distributes notions of suspiciousness (alert levels) about individual Internet hosts, as expressed by a set of participating Collaboration-enabled Detection Systems (CDS) at different autonomous systems (AS). It is built upon five main design principles: (i) each CDS must contribute in order to receive information (fair sharing); (ii) no

J. Domingo-Pascual, Y. Shavitt, and S. Uhlig (Eds.): TMA 2011, LNCS 6613, pp. 189–192, 2011.
© Springer-Verlag Berlin Heidelberg 2011

restrictions should be imposed on the nature or the configuration of a CDS, as long as it is able to interface with COMINDIS; (iii) no privacy-sensitive information should ever leave an AS; (iv) the detection sensitivity can be dynamically adapted to consider local network conditions and available resources; (v) each operator should be able to draw his own conclusions locally from the collaborative feedback.

Gray Area. A central idea of this approach is the definition of a so-called *Gray Area* of reasoning, which is defined by an upper (UTH) and a lower threshold (LTH) over the suspiciousness range. UTH is identical to the threshold of a standalone detection system. All events exceeding UTH are considered suspicious enough to be immediately investigated, without any further feedback from the collaborative system, but can still be reported to it. Events below LTH are too less suspicious to even be sent to the collaborative system, as the mere processing and communication cost exceed the operator's resources. Events in the gray area between the UTH and LTH provide enough evidence to be assigned an alert level, and trigger requests for additional information. It is important to note that UTH and LTH can be set arbitrarily by the individual operator, which can also change them over time in order to adapt to current network activity (e.g. at night, when less traffic is seen, a lower UTH can increase sensitivity).

After having received the collaborative feedback, a CDS must eventually decide on whether to classify an event as "good" or "bad". We call the corresponding COMINDIS component the Final Evaluation Function (FEF). FEF takes as input the collected alert levels for a specific host, plus other information that the operator considers important, as e.g.: the age of a report, the level of trust in the remote CDSes or the currently available resources for incident investigation. The output of FEF is evaluated by using a Final Threshold (FTH), which can be tuned by the individual operator to match the local requirements.

Architecture. The envisioned system is closed and assumes that CDSes undergo some initial key exchange procedure to keep their data exchanges secure. A CDS is by definition any monitoring device that can express an opinion about a certain Internet host in the form of a suspiciousness score, depending on the traffic that it observed. A central mediator component (the *Report Manager* (RM)) receives, stores, and distributes the reports of the individual CDSes. There is no further centralized analysis foreseen. Figure 1 shows an overview of the proposed architecture. In this example, one operator owns a powerful analyzer for detecting spammers, while another operator is able to detect DDoS attackers. Both can express their opinions about the suspiciousness regarding a specific host, even if their analysis techniques are completely different, and without disclosing any traffic data. Individually derived alert levels (here: 0.4 and 0.7) are communicated to the Report Manager. As soon as a CDS commits a report about *srcIP* at RM, it is entitled to receive all *future* and all *past* reports regarding srcIP, until the contributed report is flushed from the database.

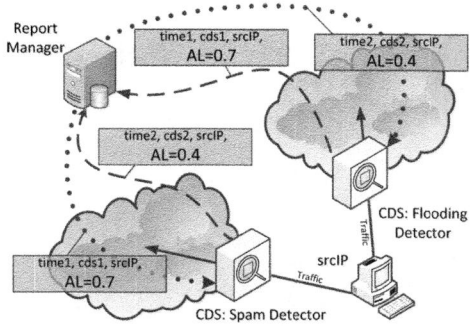

Fig. 1. Overview of the COMINDIS architecture

Information Exchange. Each report contains a timestamp, the identifier of the reporting CDS, the IP address of the suspicious host, and a normalized alert level. COMINDIS requires CDSes to submit a report before they get notified asynchronously about other reports concerning the same Internet host. Together with report rate limitation (enforced by RM), this incentive-driven model ensures fair sharing of reports. Therefore, operators must provide information about those hosts they are most interested in (i.e., with the highest local suspiciousness score), in order to not exceed their report rate quota.

Initial Results: For testing our system under realistic traffic conditions, we used a well-documented 45-minutes long trace from the OpenPacket repository that we replayed with the *tcpreplay* tool. Using the provided network architecture diagram, we assigned the network hosts to four fictive operator networks [A,B,C,D] and analyzed the inbound traffic. The trace contained 564 unique traffic sources. For each network we set the thresholds defining the *Gray Area* so that all reported events fall in there. Then, we flooded the networks using five instances of TFN2k tool for DDoS attacks, and tuned the attack rates so that the corresponding reports fall in the *Gray area*. First, each TFN2k instance attacked one host in networks [A,B,C]. After 15 minutes we shifted the attack to networks [B,C,D], to investigate the effects of collaboration over time. Reports older than 15 minutes were not taken into account.

Figure 2 shows the results for networks A and B. We evaluate the performance of the individual standalone probes (*so*) compared to COMINDIS (*coop*), and show for both the average alert levels (*avg*) for known attackers (true positives – *tp*) and legitimate hosts (false positives – *fp*), over time. For example, the *coop_avg_tp* value shows the average alert level of the attackers, as reported by the collaborative system. Note that the gain of using COMINDIS is different for each network, as the received feedback depends on the submitted reports and therefore on the individually observed traffic as well as on the CDS' threshold settings. We set FTH so to detect 100% of the attackers (see dashed annotated line in Fig. 2), and compared the standalone vs. the collaborative system performance. The annotations (x, y, z) should be interpreted as follows: x is the number of false

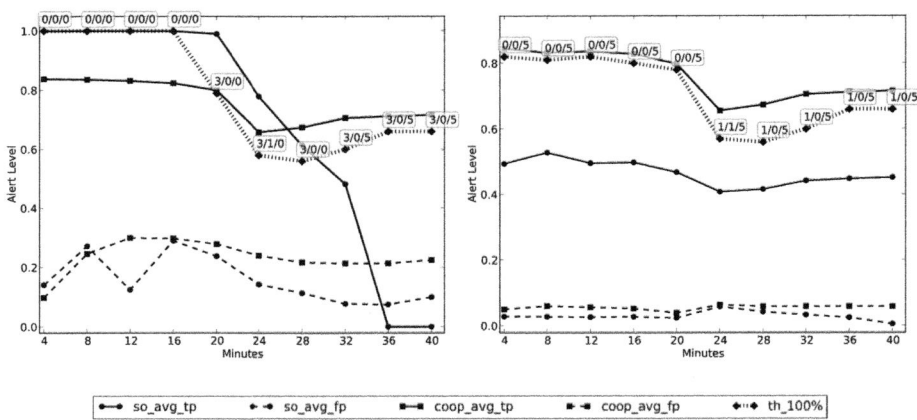

Fig. 2. Experimental results

positives that are reported by COMINDIS as a consequence of choosing this threshold. y (z) is the number of false (true) positives that would be reported by the standalone probe, if it was operated at the same threshold.

During the attack, A experiences no change in detection performance. In this scenario it is clearly a pure contributor to COMINDIS, as it is able to detect autonomously all events with high accuracy. However, note that the attackers can still be detected after 15 minutes (when the attack against A ended), as CDS A keeps receiving reports from RM. B benefits from the system and reports only at most one false positive when the standalone system would miss all five attackers.

Acknowledgements. The research leading to these results has received funding from the European Union's Seventh Framework Programme ([FP7/2007-2013]) under grant agreement n° 257315.

References

1. Allman, M., Blanton, E., Paxson, V.: An architecture for developing behavioral history. In: Proc. of Steps to Reducing Unwanted Traffic, Cambridge, MA (2005)
2. Burkhart, M., Strasser, M., Many, D., Dimitropoulos, X.: SEPIA: Privacy-Preserving aggregation of Multi-Domain network events and statistics. In: USENIX Security Symposium (2010)

Dynamic Monitoring of Dark IP Address Space (Poster)*

Iasonas Polakis, Georgios Kontaxis, Sotiris Ioannidis, and Evangelos P. Markatos

Institute of Computer Science, Foundation for Research and Technology, Hellas
{polakis,kondax,sotiris,markatos}@ics.forth.gr

Abstract. A number of security-related research topics are based on the monitoring of dark IP address space. Unfortunately there is large administrative overhead associated with the dynamic assignment of a specific subnet for monitoring purposes, such as the deployment of a honeypot farm or a distributed intrusion detection system. In this paper, we propose a system that enables the dynamic allocation of an unadvertised IP address subnet for use by a monitoring sensor. The system dynamically selects network subnets that have been allocated to the organization but are not being advertised, advertises them, and subsequently forwards all received traffic destined to the selected subnet to a monitoring sensor.

1 Introduction

An important area of Internet research focuses on monitoring of computer networks. Particularly in security-related fields, unused IP address subnets are considered a valuable resource which can enable the collection and analysis of attack traffic. By deploying intrusion detection systems, monitoring systems or honeypot farms (all of which will be referred to as *sensors*), researchers can collect vast amounts of traffic data. As a result, it is very common for researchers across organizations to collaborate and "donate" IP address subnets that are not in use.

The deployment of sensors entails a high administrative overhead. This procedure consists of several phases. First, if the subnet has been allocated but is not being advertised via some routing protocol, it must be, so that it becomes reachable by neighboring networks and may receive attack traffic. Next, routing tables inside the internal network must be altered to forward all traffic destined to the subnet to a specific machine, in our case the *sensor*. Finally, in cases where the subnet must be revoked or substituted with another one, all changes must be done manually which is prone to human error.

To facilitate the dynamic handling of subnets for the deployment of network monitoring sensors, we propose a system that will automate the procedure. The system will be aware of the dynamic routing protocol that is in effect both inter-domain and intra-domain, select the unused subnet that will be monitored, advertise it to the appropriate neighboring routers, and update routing tables so as to forward incoming traffic to the sensor. It will also detect when previously unused subnets are advertised, therefore claimed for use by the organization, and automatically release the specific subnet in respect to the privacy of its users.

---

* This work was supported in part by the project SysSec funded in part by the European Commission, under Grant Agreement Number 257007.

J. Domingo-Pascual, Y. Shavitt, and S. Uhlig (Eds.): TMA 2011, LNCS 6613, pp. 193–196, 2011.
© Springer-Verlag Berlin Heidelberg 2011

Our system, consists of 3 different components. The core component of the system is responsible for managing the other components and keeping an overview of the subnets used by the organization. It also keeps a history of all subnets advertised by an organization's router and decides which unadvertised subnets can be allocated for monitoring. The second component receives commands from the core component that instructs it to start advertising subnets that are not being advertised by the organization and which will be forwarded to the monitoring sensor. It also detects when monitored subnets are advertised by the organization, and informs the core. The final component receives commands from the core and dynamically alters the organization's internal routing tables and adds or removes entries that forward traffic to the monitoring sensor.

2   Related Work

Quagga[1] is a free, open-source network routing software suite providing implementations of protocols such as OSPF, RIP and BGP for Unix platforms. A system with Quagga installed acts as a dedicated software router. By supporting both OSPF and BGP it may be used for inter-domain as well as intra-domain routing. Quagga exchanges routing information with other, neighboring routers using routing protocols. It uses this information to update the routing table of the Unix kernel.

MAPI [5] offers an API for generic passive network monitoring based on the *network flow* abstraction, which enables users to communicate their needs to the underlying traffic monitoring platform. Moreover, MAPI offers the capability of distributed network monitoring using multiple remote monitoring sensors, and supports several different hardware platforms. DECON [3] is a decentralized and scalable coordination system that aims to solve the problem of flow assignment among a set of monitoring sensors. A peer-to-peer overlay network receives reports from all sensors that see a specific flow, and subsequently assigns the flow to one of the sensors based on a first-fit or best-fit strategy. Luca Deri [1] proposes a new dynamic packet filtering technique which overcomes the limitation of BPF by allowing users to specify several filters simultaneously and add or remove filters dynamically without any reconfigurations or downtime. Another dynamic packet filter, Swift [6], three orders of magnitude faster than BPF aims at in-place filter updating.

The Honey@home [4] architecture relies on communities of regular users and organizations installing a lightweight, traffic redirector that monitors unused IP addresses and ports. Similarly, Collapsar [2] deploys traffic redirectors in multiple network domains and examines the redirected traffic in a centralized farm of honeypots. In both cases, deployed probes require some form of initialization regarding the address space they monitor and are unable to adapt to changes in the network schema.

3   Architecture

The idea behind our system is to enable administrators to deploy a *plug-and-play* network monitoring solution. The network advertises routing prefixes for the subnets that are live, i.e., subnets that contain devices that need to communicate with the Internet.

---

[1] http://www.quagga.net/

Such advertisement already takes place using the existing infrastructure. With the addition of our monitoring solution, the administrator is required to do nothing more than connect their components to the network. The only case were some minor action is required is if the network advertises the dark subnets as well. While this is not the usual case, if so, the administrator will have to stop it, and advertise only the part of the network actually being used.

Configuration file. This file contains all the information regarding the specific organization where the sensor will be deployed. It contains the IP address range that has been allocated to the organization. Furthermore, information regarding the sensor is contained, so our system may be able to forward all traffic destined to the monitored subnet back to the sensor by dynamically altering the necessary routing tables.

Allocator component. This component is responsible for the management of all the components and dynamically changes the system's behavior based on messages received from the other components. First of all, the Allocator parses the configuration file and splits the organization's IP address range into subnets of the specified size. Based on the available subnets, the core component instructs the Advertiser component to monitor all announcements by the organization's border router. Based on that information the component can infer which subnets are dark and select one for monitoring. Then, the Advertiser is instructed to start advertising the selected subnet, and the Updater to update the routing tables and forward all traffic destined to the subnet back to the sensor. In certain cases, the system will have to dynamically change the monitored subnet. In those cases, the Allocator instructs the remaining components to stop all actions and remove all entries concerning the previous subnet. Based on the announcement history, a new subnet is selected for monitoring and all components are instructed accordingly.

Advertiser component. This component monitors all announcements by the organization's border router and keeps a history of all advertised subnets which it sends to the Allocator component. In order to do so, the administrator must designate the host running this component as a "neighboring" router (or peer) to the border router's configuration. This must be done once during the deployment phase of our system. When the appropriate instruction is received from the Allocator, the Advertiser starts advertising the selected subnet. As the host running this component does not reside at the border of the network, the border router must be configured to accept and propagate advertisements coming from our system. This acts both as a fail-safe and an assurance towards network administrators that they can be aware, control and block network updates pushed by our system. If at any moment the subnet is advertised from the border router itself, meaning that the subnet has been selected by the organization to be used, the Advertiser ceases to advertise the subnet and informs the Allocator.

Updater component. This component is responsible for updating and maintaining the routing table entries concerning the monitored subnets, in the Unix kernel of the host routing the advertised subnets of our system. That host is designated as the responsible router for a given prefix by the Advertiser, during its advertisements. The Allocator instructs the Updater to create new entries that will forward all traffic arriving at the routing host, destined to the selected subnet, back to the monitoring sensor. When the

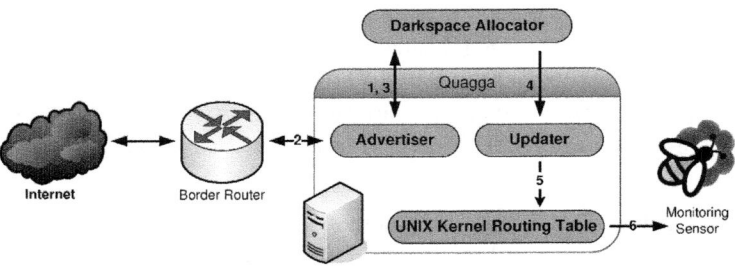

Fig. 1. System Architecture

system dynamically shifts from one subnet to another, the Updater is instructed to clear all entries regarding the previous subnet and add entries for the new subnet.

We can see a depiction of the system's architecture in Figure 1. The Allocator instructs (1) the Advertiser to log all subnet announcements from the border router (2) which are used to update the announcement history (3) and select the subnet that will be monitored. The Updater is then instructed (4) to alter the routing tables and add entries to the routing tables for the selected subnet (5) so the appropriate traffic can be forwarded to a sensor (6), such as Honey@home.

4   Conclusion

We presented the design of a system that enables the dynamic allocation of dark address space for monitoring purposes. Our system aims to facilitate organizations that want to donate IP address space for monitoring purposes, and allows the automatic handling of unadvertised IP address subnets. This is work in progress, and we are currently in the process of implementing a prototype of our system.

References

1. Deri, L.: High-speed dynamic packet filtering. Journal of Network and Systems Management 15(3), 401–415 (2007)
2. Jiang, X., Xu, D.: Collapsar: A VM-Based Architecture for Network Attack Detention Center. In: Proceedings of the 13th USENIX Security Sumposium (2004)
3. Di Pietro, A., Huici, F., Costantini, D., Niccolini, S.: Decon: Decentralized coordination for large-scale flow monitoring. In: IEEE Conference on Computer Communications, INFOCOM (2010)
4. Anagnostakis, K., Antonatos, S., Markatos, E.P.: Honey@home: A new approach to large-scale threat monitoring. In: The Proceedings of the 5th ACM Workshop on Recurring Malcode, WORM (2007)
5. Trimintzios, P., Polychronakis, M., Papadogiannakis, A., Foukarakis, M., Markatos, E., Øslebø, A.: DiMAPI: An application programming interface for distributed network monitoring. In: Proceedings of the 10th IEEE/IFIP Network Operations and Management Symposium, NOMS (2006)
6. Wu, Z., Xie, M., Wang, H.: Swift: a fast dynamic packet filter. In: Proceedings of the 5th USENIX Symposium on Networked Systems Design and Implementation, NSDI 2008 (2008)

Author Index

GPSR Compliance

The European Union's (EU) General Product Safety Regulation (GPSR)
is a set of rules that requires consumer products to be safe and our
obligations to ensure this.

If you have any concerns about our products, you can contact us on
ProductSafety@springernature.com

In case Publisher is established outside the EU, the EU authorized
representative is:

Springer Nature Customer Service Center GmbH
Europaplatz 3
69115 Heidelberg, Germany

Batch number: 09490872

Printed by Printforce, the Netherlands